# Single Cell Diagnostics

# METHODS IN MOLECULAR MEDICINE™

## John M. Walker, SERIES EDITOR

METHODS IN MOLECULAR MEDICINE™

# Single Cell Diagnostics

*Methods and Protocols*

Edited by

## Alan Thornhill, PhD, HCLD

*The London Bridge Fertility,*
*Gynecology and Genetics Centre,*
*London, UK*

HUMANA PRESS ✳ TOTOWA, NEW JERSEY

© 2007 Humana Press Inc.
999 Riverview Drive, Suite 208
Totowa, New Jersey 07512

**www.humanapress.com**

This publication is printed on acid-free paper. ∞
ANSI Z39.48-1984 (American Standards Institute) Permanence of Paper for Printed Library Materials.

Production Editor: Tara Bugg

Cover design by Nancy K. Fallatt.

Cover illustration: Single binucleated human blastomere after embryo biopsy. Courtesy of Jon Tayor, The Bridge Fertility, Gynaecology and Genetics Centre, London, UK.

For additional copies, pricing for bulk purchases, and/or information about other Humana titles, contact Humana at the above address or at any of the following numbers: Tel.: 973-256-1699; Fax: 973-256-8341; E-mail: orders@humanapr.com; or visit our Website: www.humanapress.com.

Printed in the United States of America. 10 9 8 7 6 5 4 3 2 1

eISBN: 978-1-59745-298-4

Library of Congress Cataloging in Publication Data

Single cell diagnostics : methods and protocols / edited by Alan Thornhill.
    p. ; cm. -- (Methods in molecular medicine ; 132)
  Includes bibliographical references and index.
  ISBN 1-58829-578-8 (alk. paper)
 1. Cytogenetics. 2. Molecular diagnosis. 3. Genetic screening. 4. Infertility. I. Thornhill, Alan. II. Series.
  [DNLM: 1. Cytogenetic Analysis. 2. Molecular Diagnostic
Techniques. 3. Infertility--therapy. 4. Polymerase Chain Reaction. 5. Preimplantation Diagnosis--methods.
W1 ME9616JM v.132 2007 / QY
95 S617 2007]
  QH430.S56 2007
  572.8--dc22

                                                                              2006013274

# Preface

Until recently, the world of diagnostics revolved around large chunks of tissue, whole blood samples, cell suspensions, or extracted DNA in microgram quantities. Given such a large starting template, problems of contamination and assay failure are rare but do still occur. When working with extremely small templates such as single cells or small groups of cells and their metabolites, these diagnostic challenges are simply magnified. Single cell diagnostics is a fast moving area even though the field has focused primarily on the analysis of embryonic blastomeres for the preimplantation diagnosis of single gene and chromosomal disorders—a relatively tiny area of diagnostics in terms of test volume. In just over a decade, there has been a diagnostic shift in the field from relatively simple singleplex PCR assays analyzed with ethidium based gel electrophoresis through sensitive fluorescent PCR involving multiplexing to expression analysis of multiple genes. Finally, the seemingly ubiquitous microarray technology has been applied to single cells for the identification of chromosome abnormalities and DNA alterations. *Single Cell Diagnostics: Methods and Protocols* is intended for clinical and research scientists as well as those providing care for couples seeking treatment for infertility or preimplantation genetic diagnosis. The aim is for all readers to extend their knowledge and expertise in analysis of single cells (whether or not that is their specific need). The majority of readers may not require assays of such exquisite sensitivity, but it has been my experience that many excellent research and diagnostic laboratories have learned useful tips from those of us struggling to salvage accurate diagnostic information from a single cell without contamination. *Single Cell Diagnostics: Methods and Protocols* starts with laser-assisted cell collection, noninvasive assessment of single cells and moves through the techniques of standard fluorescence *in situ* hybridization and polymerase chain reaction (PCR). As the reader moves through the book, the scope and complexity of each technique gradually increases as real-time quantitative PCR, isothermal whole genome amplification, comparative genomic hybridization, real-time gene expression analysis and the production of RNA and cDNA libraries are covered. The book closes with the application of customized microarrays to the study of single cells.

The future may see (1) a further shift away from preimplantation genetic diagnosis and more toward more routine diagnostic analysis in diseases such as cancer (in situations where very little tissue might be available for analysis) and (2) direct proteomic analysis and indirect analysis (via the secretome) from single cells.

**Alan Thornhill,** *PhD, HCLD*

# Contents

# Contributors

JAMES ADJAYE, PhD • *Molecular Embryology and Aging Group, Department of Vertebrate Genomics, Max Planck Institute for Molecular Genetics, Berlin, Germany*

DIANA W. BIANCHI, MD • *Department of Obstetrics and Gynecology, Tufts University School of Medicine, and Division of Genetics Tufts-New England Medical Center, Boston, MA*

DAVID CRAM, PhD • *Monash IVF, Melbourne, Australia*

MARTINE DE RYCKE, MEng, PhD • *Centre for Medical Genetics and Research Group Reproduction and Genetics, Medical School of the Dutch-Speaking Brussels Free University (Vrije Universiteit Brussel, VUB), Brussels, Belgium*

AHMAD EL-SHEIKHAH, PhD • *Harris Birthright Research Centre for Fetal Medicine, King's College Hospital, London, UK*

JOLENE FREDRICKSON, MSc • *Mayo Clinic College of Medicine Department of Laboratory Medicine and Pathology, Rochester, MN*

DAVID K. GARDNER, PhD • *Colorado Center for Reproductive Medicine, Englewood, CO*

XIN YUAN GUAN, PhD • *Department of Clinical Oncology, School of Chinese Medicine, The University of Hong Kong, Poke Fu Lam, Hong Kong*

SINUHE HAHN • *Laboratory for Prenatal Medicine, University Women's Hospital/Department of Research, Basel, Switzerland*

WOLFGANG HOLZGREVE • *Laboratory for Prenatal Medicine, University Women's Hospital/Department of Research, Basel, Switzerland*

DONG GUI HU, PhD • *Research Centre for Reproductive Health and Reproductive Health Science, Department of Obstetrics and Gynaecology, The Queen Elizabeth Hospital, The University of Adelaide, Woodville, SA*

NICOLE HUSSEY, PhD • *Research Centre for Reproductive Health and Reproductive Health Science, Department of Obstetrics and Gynaecology, The Queen Elizabeth Hospital, The University of Adelaide, Woodville, SA*

LONG JIN, PhD • *Department of Laboratory Medicine and Pathology, Mayo Clinic Medical College, Rochester, MN*

KIRBY L. JOHNSON, PhD • *Department of Obstetrics and Gynecology, Tufts University School of Medicine, and Division of Genetics Tufts-New England Medical Center, Boston, MA*

RICARDO V. LLOYD, MD, PhD • *Department of Laboratory Medicine and Pathology, Mayo Clinic Medical College, Rochester, MN*

JILL L. MARON, MD, MPH • *Department of Obstetrics and Gynecology, Tufts University School of Medicine, and Division of Genetics Tufts-New England Medical Center, Boston, MA*

NOBUKI NAKAMURA, PhD • *Department of Laboratory Medicine and Pathology, Mayo Clinic Medical College, Rochester, MN*

KYPROS NICOLAIDES • *Harris Birthright Research Centre for Fetal Medicine, King's College Hospital, London, UK*

CAROLINE MACKIE OGILVIE, DPhil • *Research and Development, Guy's & St Thomas' NHS Foundation Trust Centre for PGD, and Cytogenetics Department, Guy's Hospital, London, UK*

KENNETH E. PIERCE, PhD • *Department of Biology, MS-008, Brandeis University, Waltham, MA*

XIANG QIAN, PhD • *Department of Laboratory Medicine and Pathology, Mayo Clinic Medical College, Rochester, MN*

KATHARINA RUEBEL, PhD • *Department of Laboratory Medicine and Pathology, Mayo Clinic Medical College, Rochester, MN*

CHELSEA SALVADO, PhD • *Monash Immunology and Stem Cell Laboratories, Monash University, Melbourne, Australia*

KAREN SCHOWALTER, MSc • *Mayo Clinic College of Medicine, Department of Laboratory Medicine and Pathology, Rochester, MN*

PAUL N. SCRIVEN, PhD • *Research and Development, Guy's & St Thomas' NHS Foundation Trust Centre for PGD, and Cytogenetics Department, Guy's Hospital, London, UK*

KAREN SERMON, MD, PhD • *Centre for Medical Genetics and Research Group Reproduction and Genetics, Medical School of the Dutch-Speaking Brussels Free University (Vrije Universiteit Brussel, VUB), Brussels, Belgium*

ALAN R THORNHILL, PhD, HCLD • *The London Bridge Fertility, Gynecology and Genetics Centre, London, UK*

LUCILLE VOULLAIRE, MSc • *Murdoch Childrens Research Institute, Parkville, Victoria, Australia, Melbourne IVF, East Melbourne, Victoria, Australia*

LAWRENCE J. WANGH, PhD • *Department of Biology, MS-008, Brandeis University, Waltham, MA*

DAGAN WELLS, PhD • *Department of Obstetrics and Gynecology, Yale University Medical School, New Haven, CT*

LEEANDA WILTON, PhD • *Murdoch Childrens Research Institute, Parkville, Victoria, Australia, Melbourne IVF, East Melbourne, Victoria, Australia*

HEYU ZHANG, PhD • *Department of Laboratory Medicine and Pathology, Mayo Clinic Medical College, Rochester, MN*

BERNHARD ZIMMERMANN, PhD • *University Women's Hospital/ Department of Research, University Hospital Basel, Switzerland*

# 1

# Noninvasive Metabolic Assessment of Single Cells

## David K. Gardner

## Summary

Because metabolism is fundamental to the successful functioning of a cell, perturbations in metabolism can signal compromised viability. For the assessment of viability, analyses not only must be sensitive enough to be employed at the single cell level, but also noninvasive. Ultramicrofluorescence is a technique that fulfills both criteria. By using specially prepared constriction pipets, created on a microforge and calibrated with tritiated water, volumes in the nano- and picoliter range can be accurately delivered. A miniaturization of enzymatic analysis using a fluorescence microscope, with either photometer or charge-coupled device attachments, is then employed to quantitate biochemical reactions in such submicroliter volumes. Subsequently, samples of culture medium surrounding an individual cell can be taken and analyzed in order to determine the rate of utilization or production of various metabolites in real time.

**Key Words:** Constriction pipet; embryo; enzymatic analysis; microfluorescence; nanoliter; picoliter; viability.

## 1. Introduction

There are several ways to quantitate the metabolism of cells. However, the majority of techniques typically involve the destruction of the cell in question or the use of radiolabeled substrates. Noninvasive assays, therefore, have the evident advantage of not perturbing cell integrity or function. However, the latter approach is not without limitations regarding the type of data that can be obtained. Whereas the use of radiolabeled substrates can assist in determining the relative activity of a specific metabolic pathway, such as the use of $[5\text{-}^3\text{H}]$ glucose to measure the activity of the Embden-Meyerhof pathway (1), noninvasive assays can only measure the rate of nutrient consumption and metabolite release. Therefore, should one wish to know the activity of glycolysis in a cell, one could quantitate glucose consumption together with lactate production. Although such an approach can give an indirect measure of glycolysis, it cannot

From: *Methods in Molecular Medicine: Single Cell Diagnostics: Methods and Protocols*
Edited by: A. Thornhill © Humana Press Inc., Totowa, NJ

exclude the finite possibility that the lactate produced could have come from another source within the cell, such as from any pyruvate generated from the metabolism of specific amino acids or from pyruvate taken up from the surrounding medium, or from the utilization of endogenous glycogen. However, even with such limitations, the ability to quantitate cell metabolism using noninvasive procedures is a powerful tool.

Analysis of metabolites within cells was initially performed using a technique of enzyme cycling *(2)*, by which it was possible to amplify substrate levels up to 30,000-fold. However, this technique is complex and takes a long time for analysis. A direct approach to analyzing biological compounds in small volumes was developed by Mroz and Lechene *(3)*, who used specifically constructed micropipets capable of delivering volumes in the nano- and picoliter range to assay femtomole amounts of urea. The micropipets were employed to deliver droplets of sample and reagent onto siliconized microscope slides under mineral oil. After the reaction went to completion, the droplets were taken up in capillary tubes, which were then placed on the stage of a fluorescence microscope and the product of the assay was quantitated. In this assay, fluorescence was generated using phthaldehyde plus thioglycolic acid, which determined the amount of ammonia produced from urea.

Leese et al. *(4)* subsequently modified this elegant technique during an investigation of the nucleotide content in single mouse oocytes and embryos. Instead of using a fluorescent reagent, the analytical procedure was a miniaturized version of conventional methods of enzymatic analysis in which the nucleotides NAD(P)H are generated or consumed in coupled reactions. These pyridine nucleotides have absorption maxima at 340 nm, and when the reduced forms are excited with light in the ultraviolet (UV) range, they emit fluorescence at 459 nm and above. Rather than take the reaction mixture into capillary tubes, the fluorescence of the droplets on the microscope slides was quantitated directly. Initial problems of enzyme denaturation at the oil/aqueous interface were overcome by increasing the enzyme content of the assay cocktail. Using this system a great variety of metabolites and enzymes can be assayed, and Leese and Barton *(5)* first applied the system to the noninvasive quantification of nutrient utilization of single embryos. It is the application of this latter approach that forms the basis for this chapter.

Fluorometric assays for metabolites and enzymes (which can be detected in the surrounding medium if the plasma membrane is disrupted) are based on the generation or utilization of the reduced pyridine nucleotides, NADH and NADPH, in coupled enzymatic reactions. These nucleotides fluoresce when excited with light at 340 nm, whereas the oxidized forms, $NAD^+$ and $NADP^+$, do not. Thus, the reaction

$$\text{Pyruvate} + \text{NADH} + \text{H}^+ \xleftrightarrow{\quad\text{Lactate Dehydrogenase}\quad} \text{Lactate} + \text{NAD}^+$$

may be followed by monitoring a drop in fluorescence as NADH is converted into NAD$^+$. Under appropriate conditions (described in detail below) the level of change in fluorescence is proportional to the amount of substrate consumed in the reaction, and, therefore, the amount can be calibrated using a standard curve. The reaction conditions are set to favor the completion of a given reaction in one way or the other. For example, in the preceding reaction the equilibrium favors the formation of lactate (using a 4-(2-hydroxyethyl)-1-piperazine propane-sulfonic acid [EPPS] buffer system at pH 8.0). However, when the more basic glycine-hydrazine buffer is used, the equilibration moves to the left. In addition, any free pyruvate is converted and effectively trapped as pyruvate-hydrazone, thereby making the reaction in effect nonequilibrium. All reactions occur in nanoliter volumes so that substrate levels in the pico- and femtomole range can be measured.

Assays for substrate levels and enzyme activity rely on quantitation of the fluorescence of the reduced forms of the pyridine nucleotides (NADH, NADPH) under UV light. Therefore, the essential equipment for this type of analysis are: a mercury lamp, to excite the sample with wavelengths in the UV range (340 nm); a shutter system or optical switch, to limit the excitation and emission of the drops to be analyzed; and a photometer attachment with software that can convert the low levels of emitted light into a numerical value. The photometer output is expressed in arbitrary units that can be calibrated using standard curves. Such quantitative systems are available commercially from most microscope manufacturers (e.g., the Leica MPV system, and the SFX-2 Micro Fluorimeter and DX-1000 Optical Switch from Solamere Technology Group Salt Lake City, UT).

## 2. Materials

1. Fluorescence microscope with photometer attachment (e.g., Nikon TE 300 with Solamere Technology SFX-2 Micro Fluorimeter and DX-1000 Optical Switch).
2. Microforge (Narashige or DeFonbrune).
3. Borosilicate glass capillaries (15 mm long, 1.0-mm od, 0.78-mm id; Harvard Apparatus, Edenbridge, UK).
4. Tritiated water (5 mCi/mL; Amersham).
5. Dimethyldichlorosilane solution (Repelcote; BDH) or SIGMACOTE (Sigma).
6. Lactate dehydrogenase (EC 1.1.1.27, rabbit muscle ; Roche).
7. Hexokinase/glucose-6-phosphate dehydrogenase (EC 2.7.1.1/1.1.1.49, yeast; Roche).
8. Glutaminase (EC 3.5.1.2, beef liver; Roche).
9. Glutamate dehydrogenase (EC 1.4.1.2, beef liver; Sigma).
10. EPPS buffer: 2.25 g of EPPS, 10 mg of penicillin, 10 mg of streptomycin made up to 200 mL in water with the pH adjusted to 8.0 with 1 $M$ NaOH.
11. Glycine-hydrazine buffer: 7.5 g of glycine, 5.2 g of hydrazine, 0.2 g of EDTA in 50 mL of water with the pH adjusted to 9.0 or 9.4 with 2 $M$ NaOH.

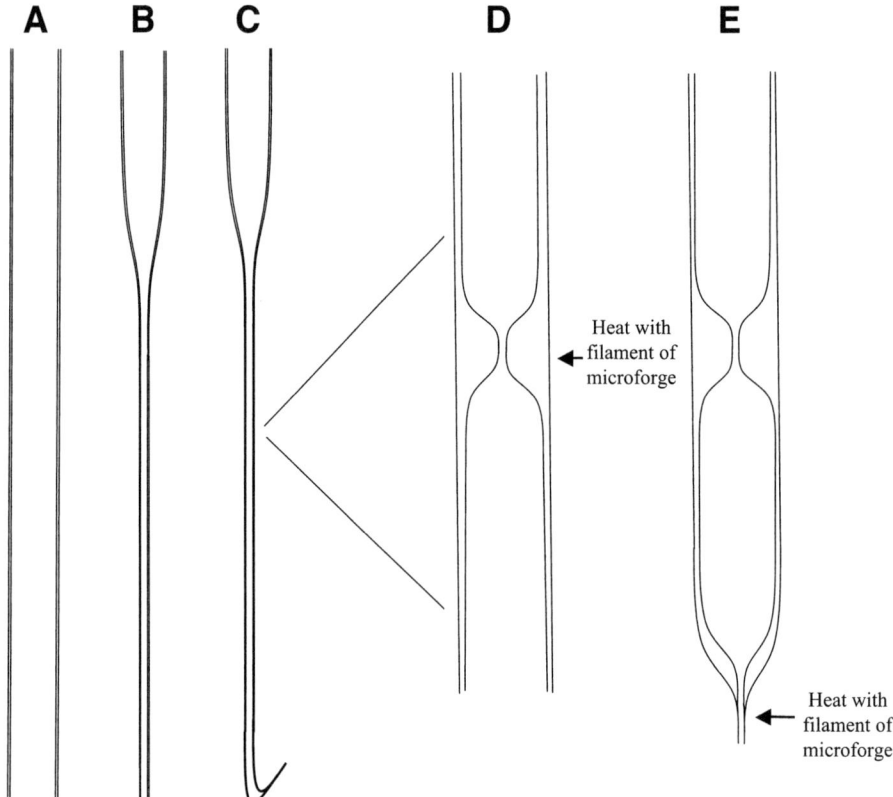

**A    B    C              D              E**

Heat with filament of microforge

Heat with filament of microforge

12. Acetate buffer: 6.8 g of sodium acetate in 100 mL of water, 2.9 mL of acetic acid in 97.1 mL of water. Mix 68 mL of sodium acetate solution with 32 mL of acetic acid solution, pH 5.0.

13. Reagents for pyruvate assay: 0.075 m$M$ NADH, 28 U of lactate dehydrogenase/mL in EPPS buffer, pH 8.0.

14. Reagents for lactate assay: 4.76 m$M$ NAD$^+$, 100 U of lactate dehydrogenase/mL, 2.6 m$M$ EDTA in glycine-hydrazine buffer, pH 9.4 .

15. Reagents for glucose assay: 3.7 m$M$ MgSO$_4$ pH·7H$_2$O, 0.6 m$M$ NADP$^+$, 0.5 m$M$ adenosine triphosphate (ATP), 0.5 m$M$ dithiothreitol, 12 U of hexokinase/mL, 6 U of glucose-6-phosphate dehydrogenase/mL in EPPS buffer, pH 8.0.

16. Reagents to convert glutamine into glutamate: 50 U of glutaminase/mL in acetate buffer, pH 5.0.

17. Reagents for glutamate assay: 1.6 m$M$ NAD$^+$, 1.0 m$M$ adenosine diphosphate (ADP), 100 U of glutamate dehydrogenase/mL in glycine-hydrazine buffer, pH 9.0.

## 3. Methods

### *3.1. Production of Micropipets*

For analysis of substrate uptake, metabolite production, or enzyme activity by single cells or embryos, the conventional assays are scaled down to occur in

Fig. 1. Production of constriction pipets for delivery of volumes in nano- and picoliter range. (**A**) A borosilicate capillary (150 mm long), with an od of 1.0 mm and an id of 0.78 mm is shown. (**B**) The capillary is heated in the middle using a small yellow flame and pulled. The capillary is then broken into two halves. Two pipets can be prepared from one capillary. The size of the pipet is determined in part by how thin the capillary is pulled. (**C**) A hook is made by melting the end of the pulled capillary using a small flame. The capillary is then mounted in a microforge with the hook end down. (**D**) Using a heated platinum filament, a constriction is made in the pulled area of the capillary. The heated filament is placed as close to the capillary without touching and the heat increased until the glass implodes within the capillary, forming a constriction. (**E**) A small weight is placed on the hook, such as half of a small paper clip. The heating element is then moved down the capillary and the heat reapplied to the capillary. The distance between the initial constriction and where the capillary is melted for the second time will in part determine the volume of the pipet (the other factor being the inner diameter of the pulled capillary). As the glass melts, rather than implode, the glass capillary is pulled to a tip as the weight on the hook pulls on the capillary. The sides on the capillary are not allowed to touch. Rather, heat is discontinued once the capillary has been drawn to a small constriction. Less heat is applied at this time than when the original constriction is made, because the glass is much thinner once it starts to be pulled and will quickly seal if the filament is too hot. Once the capillary has been drawn out, the glass can be broken using a pair of watchmaker forceps, by pulling on the capillary below the position of the tip. The capillary will break square, leaving an intact constriction pipet. The newly constructed pipet is next mounted in a 16-gage stainless-steel tube and fixed in place using sealing wax. The mounted pipet can then be siliconized and calibrated. Such a pipet can be reused indefinitely, provided it is cleaned between uses and resiliconized when required.

submicroliter volumes. The submicroliter volumes are manipulated by specially constructed constriction micropipets. These pipets are made from borosilicate glass capillaries (1.0-mm od, 0.78-mm id) pulled over a flame to produce an id of about 50–100 µm. The tubing is broken in half, a small hook is made on each end using a small open flame, and the tubing is then mounted on a microforge. Therefore, each capillary tube can be used to make two pipets. Using the microforge, a constriction is made in the glass by placing the heated filament close to the glass capillary. A small weight (half a paper clip) is then placed on the end of the hook of the capillary and the tip is made by again heating the glass with the microforge filament. As the glass heats up, the weight pulls the glass to form a tip. The tip is then broken with a pair of watchmaker forceps by pulling down on the capillary just below the tip. The glass will typically break square. The size of the constriction and tip control the speed and accuracy of the pipet. Pipets are mounted in 16-gage stainless steel tubing and sealed using sealing wax. **Figure 1** schematically shows the procedure for making constriction pipets. **Figure 2** shows the constriction pipet itself in detail.

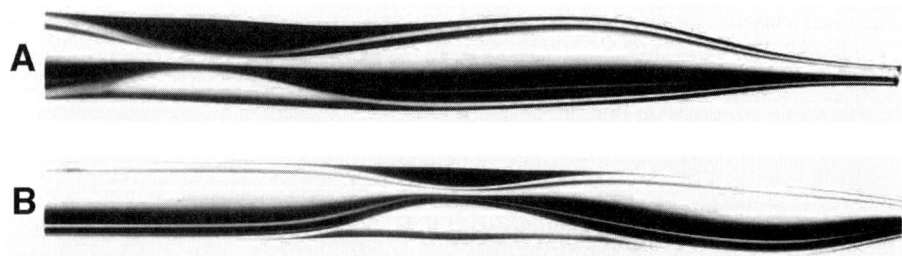

Fig. 2. **(A)** Photograph of a 5-nL constriction pipet. **(B)** Photograph of constriction within 5-nL pipet shown in (A). Note that if the constriction were any thinner than this, it would be difficult to take up samples and/or reagents. If the constriction were any larger, the pipet would be difficult to use because the sample/reagent would not stop readily at the constriction.

Filling and expelling fluids from the pipets is achieved using an air-filled syringe attached to the pipets via tubing. Prior to use, the micropipets are siliconized to prevent fluids from sticking to the small internal diameter of the pipets, thereby facilitating accurate delivery. Siliconization of the pipets can be achieved by taking up dimethyldichlorosilane solution or SIGMACOTE. The pipets are then held in a micromanipulator and the volumes manipulated under a microscope. The volume of the pipets between the tip and the constriction is calibrated using tritiated water. Stocks of the original 5 mCi/mL tritiated water are prepared as a 1 in 5 dilution and a 1 in 50,000 dilution. The 1 in 5 dilution is used to calibrate the constriction pipets. The diluted tritiated water is dispensed from the constriction pipets into scintillation vials containing an appropriate scintillation fluid. The constriction pipet is then washed and the process repeated to generate triplicate samples from each pipet to be calibrated. The 1 in 50,000 dilution is used as a standard for which known calibrated pipets in the microliter range can be used. Such microliter volumes are dispensed into scintillation vials and triplicate vials are set up. The volume of the constriction pipet can then be calculated from the disintegrations per minute value obtained following counting in a scintillation counter. Using this procedure it is possible to accurately pipet volumes in the nano- and picoliter range *(1)*. To control the constriction pipets, a micromanipulator and stereomicroscope are required. It is essential to prevent blockage in the constriction pipets (*see* **Note 1**).

### 3.2. Assays

1. The assays themselves are housed in submicroliter droplets on siliconized microscope slides under heavy mineral oil (heavy white grade; Sigma). For each assay, reagent cocktail solutions are prepared that contain a buffer and all of the cofactors and enzymes needed for the reaction. Typically, 10- to 20-nL drops (depending on the substrate to be analyzed and the size of the cell being studied) of this reagent cocktail are placed onto a siliconized slide under the mineral oil to prevent evaporation.

2. The fluorescence of each droplet is measured in turn by exposing the pyridine nucleotides in the cocktail to the UV light source. Drops are measured using a ×20 or ×40 objective and are routinely exposed for up to 0.5 s, since there is no detectable photooxidation of NADH or NADPH during this time.

3. Following this initial determination of fluorescence, sample (1–5 nL depending on the substrate to be analyzed) is added to the reagent cocktail drop. The addition of the substrate initiates the reaction. The drops on the slide are then left until the reaction has gone to completion (*see* **Note 2**).

4. The fluorescence of the drops is again determined. The change in fluorescence between the reagent cocktail drop before and after addition of the sample should be linear within the concentrations to be assessed. It is necessary to run a new set of standards on each day of an experiment to determine whether the fluorescence range is linear within the concentration of the substrate. An acceptable linear regression value is typically $R > 0.99$.

5. Once a linear standard curve has been achieved, the concentration of substrates from samples can be calculated from this standard. The reactions and assay conditions for pyruvate, lactate, glucose, and glutamine are as follows:

**Pyruvate Assay**

$$\text{Pyruvate} + \text{NADH} + \text{H}^+ \xrightarrow{\textit{Lactate Dehydrogenase}} \text{Lactate} + \text{NAD}^+$$

**Lactate Assay**

$$\text{Lactate} + \text{NAD}^+ \xrightarrow{\textit{Lactate Dehydrogenase}} \text{Pyruvate} + \text{NADH} + \text{H}^+$$

**Glucose Assay**

$$\text{Glucose} + \text{ATP} \xrightarrow{\textit{Hexokinase}} \text{Glucose-6-phosphate} + \text{ADP}$$

$$\text{Glucose-6-phosphate} + \text{NADP}^+ \xrightarrow{\textit{Glucose-6-phosphate Dehydrogenase}} \text{6-phosphogluconate} + \text{NADPH} + \text{H}^+$$

**Glutamine Assay**

Step 1

$$\text{Glutamine} + \text{H}_2\text{O} \xrightarrow{\textit{Glutaminase}} \text{Glutamate} + \text{NH}_3$$

Step 2

$$\text{Glutamate} + \text{H}_2\text{O} + \text{NAD}^+ \xrightarrow{\textit{Glutamate Dehydrogenase}} \alpha\text{-ketoglutarate} + \text{NH}_4^+ + \text{NADH}$$

## 3.3. Measurement of Nutrient Uptake by Individual Cells

Cells are incubated in a medium containing the substrates for measurement at levels typically between 0.1 and 1.0 m$M$. This medium can be either HEPES or 3-($N$-morpholino) propanesulfonic acid (MOPS) buffered medium in which the incubation can take place on a calibrated heating stage at 37°C, or it can be bicarbonate buffered in which the incubation must occur in a $CO_2$ incubator (*see* **Note 3**). Cells or embryos are typically incubated in this medium for 1–4 h depending on the volume of the drop, size of the cell, or stage of an embryo to be analyzed. Whenever possible, linear rates of uptakes should be determined by taking serial measurements over the course of the incubation. Serial measurements ensure that there are no alterations in substrate utilization as a result of changes in the culture conditions.

For incubation of cells or embryos, drops of metabolic incubation medium are placed under oil in a sterile Petri dish (*see* **Note 4**). For the mouse, it is typical to incubate embryos in drops of medium in the 20- to 50-nL range, whereas for domestic animal embryos or human embryos volumes of 200 nL to 1 µL are used. Embryos are washed well in the defined incubation medium to ensure that there is no carryover of substrates to the small incubation drops. The embryos are then picked up in a very small volume using a pulled pipet just larger than the embryos themselves. This will ensure that there is minimal carryover of extra volume into the drops. The volume that is carried over into the drops can be determined using tritiated water *(6)*.

For determination of linear rates of uptake or production, serial samples are taken at 30- to 45-min intervals. A minimum of three readings is required to determine linear rates. For end-point determinations, 3 to 4 control drops of medium alone (containing no embryos or cells) should be included in the incubation to determine the exact amounts of nutrients that were present in the medium and control for any breakdown that may have occurred during the incubation period. If the samples are not to be analyzed immediately, the cell or embryo can be removed from the drops at the end of the incubation and the medium taken up in 1- to 5-µL capillary tubes surrounded by oil on each end to avoid evaporation. Such capillary tubes can then be placed in plastic insemination straws and stored at –80°C. There is negligible breakdown of most substrates during storage for several weeks (<1%) under these conditions.

## 4. Notes

1. Should a constriction pipet become blocked, which is typical after intensive use, owing to the buildup of protein within the narrow lumen, it can be cleaned using dilute acid (e.g., 0.1 $M$ HCl). After soaking the pipet in dilute acid, it must then be resiliconized before reuse.
2. Time for the reaction to reach completion can vary between 3 min and 1 h, depending on the substrate being analyzed (this must be determined for each assay for a substrate by taking readings over time to determine when the reaction has reached completion).
3. If the medium is to be used outside of a $CO_2$ environment, replace 23 m$M$ bicarbonate with either HEPES or MOPS *(7)*.

4. It is preferable to incubate cells or embryos in dishes with low sides or the lid of a 35-mm Petri dish. If using a dish lid, first ensure that the lid is not embryo toxic.

## Acknowledgments

I am grateful to Katie Stephens for assisting with the production of this chapter and to Kim Preis and Dr. Mark Johnson for their comments on the manuscript.

## References

1. Rieger, D. (2004) Metabolic pathway activity, in *A Laboratory Guide to the Mammalian Embryo* (Gardner, D. K., Lane, M., and Watson, A. J., eds.), Oxford University Press, Oxford, NY, pp. 154–164.
2. Lowry, O. H. and Passoneau, J. V. (1972) *A Flexible System of Enzymatic Analysis,* Academic, New York.
3. Mroz, E. A. and Lechene, C. (1980) Fluorescence analysis of picoliter samples. *Anal. Biochem.* **102,** 90–96.
4. Leese, H. J., Biggers, J. D., Mroz, E. A., and Lechene, C. (1984) Nucleotides in a single mammalian ovum or preimplantation embryo. *Anal. Biochem.* **140,** 443–448.
5. Leese, H. J. and Barton, A. M. (1984) Pyruvate and glucose uptake by mouse ova and preimplantation embryos. *J. Reprod. Fertil.* **72,** 9–13.
6. Gardner, D. K. and Leese, H. J. (1986) Noninvasive measurement of nutrient uptake by single cultured preimplantation mouse embryos. *Human Reprod.* **1,** 25–27.
7. Lane, M. and Gardner, D. K. (2004) Preparation of gametes, in vitro maturation, in vitro fertilization, and embryo recovery and transfer, in *A Laboratory Guide to the Mammalian Embryo* (Gardner, D. K., Lane, M., and Watson, A. J., eds.), Oxford University Press, Oxford, NY, pp. 24–40.

# 2

# Laser Capture Microdissection for Analysis of Single Cells

## Nobuki Nakamura, Katharina Ruebel, Long Jin, Xiang Qian, Heyu Zhang, and Ricardo V. Lloyd

## Summary

Laser capture microdissection (LCM) can be used to obtain single cells or a homogeneous population of cells for molecular analysis. This approach becomes even more powerful when it is combined with immunocytochemical staining using specific antibodies to label the cells of interest before LCM (referred to as immuno-LCM). These techniques have been applied in our laboratory to the analysis of pituitary cells from dissociated tissues and from cultured populations of heterogeneous pituitary, thyroid, and carcinoid tumor cells, as well as for the analysis of single cells in various sarcomas. When combined with reverse transcriptase polymerase chain reaction (RT-PCR) and Southern blot analysis, the sensitivity of this method is increased, allowing the reproducible analysis of gene expression from 1 to 10 cells. These methods show the utility of immuno-LCM as well as LCM combined with RT-PCR for cellular and molecular studies of gene expression.

**Key Words:** Laser capture microdissection; pituitary cells; reverse transcriptase polymerase chain reaction; cell culture; immunocytochemistry; Southern hybridization.

## 1. Introduction

Collection of homogeneous populations of cells for biological and molecular analysis was a difficult challenge before the development of laser capture microdissection (LCM) *(1–3)*. Analysis of single cells and small groups of homogeneous cells has been facilitated by the use of LCM, especially in the study of complex disease processes *(4–15)*. In our laboratory, we have performed analyses of pituitary, thyroid, carcinoid, and sarcoma cells using LCM followed by reverse transcriptase polymerase chain reaction (RT-PCR) and Southern hybridization. We have also analyzed pituitary cells by first performing immunocytochemical staining previous to LCM (immuno-LCM). In this

From: *Methods in Molecular Medicine: Single Cell Diagnostics: Methods and Protocols*
Edited by: A. Thornhill © Humana Press Inc., Totowa, NJ

combined procedure, antibodies specific for certain cell types are used to attain a higher degree of specificity and selectivity of the individual cells or cell populations being studied *(3,4)*. Immuno-LCM has been used to obtain pure populations of pituitary cells producing prolactin (PRL) and from folliculostellate cells, which express abundant S100 protein *(8)*. After immuno-LCM or LCM, total RNA is extracted and molecular analysis is performed by RT-PCR. The PCR-amplified products may then be analyzed by agarose gel electrophoresis and subsequently by Southern hybridization, which increases the sensitivity of the method. This approach allows the analysis of gene expression in one cell or in a small group of homogeneous cells.

## 2. Materials

1. GH-3 cells maintained in culture.
2. Thyroid cells maintained in culture.
3. Anterior pituitary cells from 60- to 90-d-old Wistar-Furth rats.
4. Frozen sections of sarcoma tissue such as Ewing's sarcoma, desmoplastic small round cell tumor, alveolar rhabdomyosarcoma, and synovial sarcoma.
5. Dulbecco's modified Eagle's medium with 15% horse serum, 2.5% fetal calf serum, and 1% antibiotic/antimycotic.
6. TRIzol (Invitrogen-Life Technologies, Carlsbad, CA).
7. Glycogen (Sigma-Aldrich, St. Louis, MO).
8. Chloroform.
9. Isopropanol.
10. Ethanol.
11. Diethylpyrocarbonate (DEPC) (Sigma-Aldrich).
12. Prostar First Strand RT-PCR Kit (Stratagene, La Jolla, CA).
13. Agarose gel.
14. Ethidium bromide.
15. Nylon membrane.
16. Standard saline citrate (SSC).
17. X-ray film (XOMAT AR Film; Kodak, Rochester, NY).
18. Easy-A High-Fidelity Cloning Enzyme, and PCR primers.
19. Trypsin (Invitrogen-Life Technologies).
20. Hematoxylin (Sigma-Aldrich).
21. Diaminobenzidine (Sigma-Aldrich).
22. Xylene (Sigma-Aldrich).

## 3. Methods

### 3.1. Preparation of Single Cells from Tissues and Immunostaining

1. Dissociate pituitary cells with 0.25% trypsin (Invitrogen). Attach dispersed cells to uncoated glass slides using between $1 \times 10^3$ and $1 \times 10^4$ cells/slide prepared by cytocentrifugation (*see* **Note 1**).
2. Fix cells in 100% ethanol for 5 min or paraformaldehyde for 20 min, air-dry for 5 min, and then use for immunocytochemistry and LCM.

3. Conduct immunocytochemistry with specific antibodies to pituitary hormones obtained from the National Pituitary Agency, Baltimore, MD. Negative control for immunostaining consists of substituting normal rabbit serum or mouse serum for the primary antibody, which should result in no staining of cells (*see* **Note 2**).
4. Lightly counterstain the cells with hematoxylin and place in 3% glycerol and RNase-free water for 20 min to facilitate cell detachment during LCM.
5. Dehydrate the slides with 95 and 100% ethanol.
6. Immerse the cells in xylene for 10 min and air-dry at room temperature prior to LCM.
7. During immunostaining, add 400 µL of RNasin to all solutions to decrease RNase contamination.
8. Incubate the slides with the primary antibody for 2–20 min, then the second antibody for 20 min followed by an avidin-biotin-peroxidase reaction for 20 min, then diaminobenzidine for 1 min (*see* **Note 3**).
9. Perform LCM with a PixCell II Laser Capture Microdissection System (Arcturus, Mountain View, CA).
10. Capture the immunopositive cells directly onto a thermoplastic polymer (ethylene vinyl acetate) film-coated cap. Use an infrared (IR) laser with 60 mW of laser power and a laser beam with a 7.5- –15-µm diameter to capture cells of interest. The laser melts the ethylene vinyl acetate from the plastic film directly onto the target cell, embedding the captured cells (**Fig. 1**).
11. After transfer to the cap, use the samples for RNA or DNA analysis (*see* **Notes 3–5**).

### 3.2. Operating PixCell II System for LCM

1. Turn on the power strip located behind the computer monitor. Also turn on the power for the monitor.
2. On the computer screen, click the mouse on "shortcut to ARC100."
3. When the instrument serial number is given, click "continue."
4. Select the name or enter a new name and click "acquire data."
5. Highlight the study or enter a new study and enter "select."
6. Enter the slide number, spot size, cap lot, and signet (usually 10 µm, but this may vary depending on the slides cut), and click "continue."
7. Enter a laser power of approx 45 mW and a pulse of approx 55 ms (to start, select lens C).
8. Place the slides on a microscope, and visualizing the monitor, find an appropriate starting area that would be easy to find after finishing the microdissection.
9. Click on "before" to obtain a picture of the area selected.
10. Insert a roll of ARCTURUS caps with transfer film onto the slot on the right side of the microscope.
11. If the laser power is <60 mW, to place the optic beam, adjust the piece without the filter in the indentation on the end of the placement arm. If the laser power chosen is >60 mW, to place the beam, adjust the piece with the filter on the arm.
12. Use a placement arm, pick up a cap, move the arm all the way over to the left (this will put it directly above the slide), and gently release it so that the cap slowly drops onto the slide.

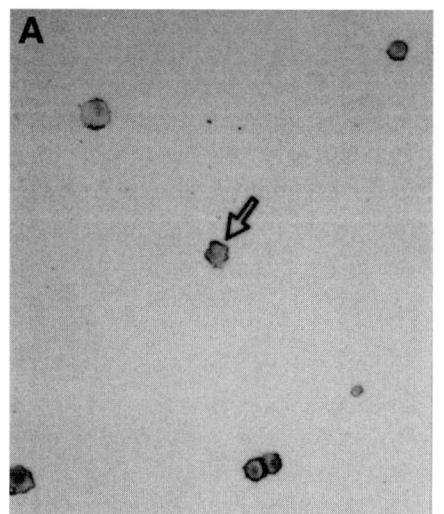
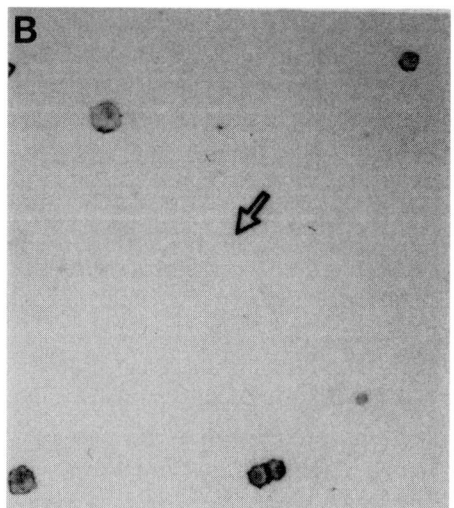
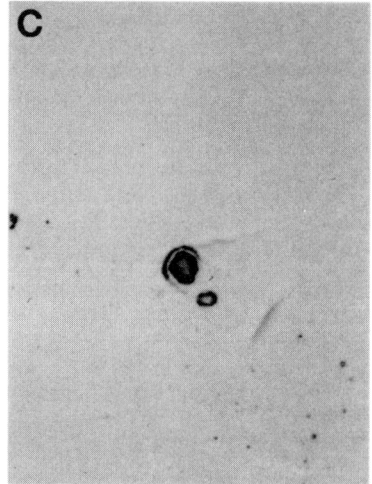

13. Pick up the white cord with the red button on the end, and press the button to get a laser pulse. A dark circle will appear around the area if it is "melted."

14. Use the joystick located to the bottom right of the microscope to move the slide around and get pulses in different areas.

15. After finishing the microdissection, use the arm to pick up the cap very gently so as not to pick up surrounding tissues, move it over to the right, and place it on a sterile 0.5-mL microcentrifuge tube.

16. Find the original starting area and click on "after" to get an image of the completed microdissection, and look at the slide under a microscope to make sure that most of the areas have been transferred.

17. Click "done" and then "exit" or "continue."

Fig. 1. LCM of cells from normal rat pituitary PRL cells. Normal rat anterior pituitaries were dissociated with 0.25% trypsin. The dispersed rat pituitary cells were attached to uncoated glass slides ($1 \times 10^3$ to $1 \times 10^4$ cells/slide) by cytocentrifugation, fixed in 100% ethanol for 5 min, air-dried for 5 min, and then used for immunocytochemistry and LCM. Immunocytochemistry for rat PRL (1/1000 dilution; obtained from the National Pituitary Agency) and adrenocorticotropic hormone (1/1000; from Dako, Carpinteria, CA) was performed. Negative controls for immunostaining consisted of substituting normal rabbit serum for the primary antibody, which resulted in no staining of the slides. The cells were lightly counterstained with hematoxylin and placed in 3% glycerol in RNase-free water for 20 min to facilitate cell detachment during LCM. The slides were then dehydrated with 95 and 100% ethanol, followed by incubation in xylene for 10 min, and air-dried at room temperature prior to LCM. During immunostaining, 400 U/mL of RNasin (Promega, Madison, WI) was added to all solutions to decrease RNase contamination. The PixCell Laser Capture Microdissection System (Arcturus) was used for LCM. The immunostained positive cells were captured directly onto a thermoplastic polymer film-coated cap by a one-step transfer method. An IR laser with 60 mW of laser power and a laser beam with a 30-μ diameter were then pulsed over the cells of interest, which melted the ethylene vinyl acetate thermoplastic film directly onto the targeted cell, embedding the captured cells. After transfer to the cap, the samples were used for RNA analysis. (**A**) An immunostained PRL cell with brown cytoplasmic staining was selected for LCM (arrow). (**B**) The arrow indicates the space from which the cell was captured and transferred. (**C**) A captured PRL cell on the cap with the transfer film is shown.

## 3.3. Extraction of RNA from Collected Cells

1. Obtain cells in a cap with transfer film using LCM.
2. Aliquot 200 μL of TRIzol reagent into a sterile 0.5-mL microcentrifuge tube and place the cap with the film on top of the tube and invert the tube (*see* **Note 4**).
3. Leave the samples with TRIzol inverted at room temperature for no more than 1 h.
4. Take off the cap and add 1 μL of glycogen and 400 μL of chloroform to each tube and vortex vigorously for 15 s.
5. Incubate the samples for 3 min at room temperature.
6. Using an Eppendorf centrifuge in a cold room, spin for 15 min at 2500*g*.
7. Transfer the aqueous phase to a new 0.5-mL tube containing 10 μL of isopropanol and vortex.
8. Incubate at room temperature for 10 min.
9. Place in a –70°C freezer for 1 h, or –20°C for at least 3 h (or overnight). Take out and allow to thaw.
10. Centrifuge for 30 min at 2500*g* in the Eppendorf centrifuge (in a cold room).
11. Very carefully discard the supernatant, because you may not see a pellet.
12. Add 200 μL of 100% isopropyl alcohol and vortex.
13. Centrifuge at 2000*g* for 5 min in the Eppendorf centrifuge.

14. Carefully pour off the supernatant, invert the tube, and air-dry for approx 5 min. Resuspend the pellet in 10 µL of DEPC water and use this directly for the RT reaction.

### 3.4. RT-PCR Analysis of Collected Cells

1. Prepare the first-strand cDNA from the total RNA using a Prostar First-Strand RT-PCR Kit (Stratagene). Perform the RT reaction in a final volume of 50 µL with 10 µL of the total RNA from the LCM transfer cells and 300 ng of oligo-dT primers at 65°C for 10 min. Then add the other reagents included in the kit, and incubate the sample at 42°C for 1 h (thermal cycler or water bath).
2. Heat at 90°C (thermal cycler or water bath) for 5 min.
3. Immediately place on ice. Use omission of the RT enzyme during the RT reaction as the negative control (*see* **Note 6**).
4. Perform PCR amplification using 30–45 cycles for the LCM samples, depending on the conditions that must be optimized for individual laboratories. Typical cycle parameters involve 40 cycles for the LCM sample (*see* **Note 5**). The annealing temperature depends on the primer but may range from 50 to 65°C.
5. After the final cycle, extend the elongation step by 10 min at 72°C.
6. Amplify a housekeeping gene such as HPRT from the same RT product and use as an internal control. Conduct analysis of other cells (including cells producing other hormones) to check for the homogeneity of the population and ensure that there are no contaminating cells.
7. Perform Southern hybridization with a 20-µL aliquot of the PCR product. Analyze the PCR product by electrophoresis in a 2% agarose gel with ethidium bromide staining.
8. Perform titration studies with different amounts of cDNA to verify that each amplification is in the linear range.
9. Transfer DNA to nylon membrane filters, and perform Southern hybridization with [33]P-labeled internal probes at 42°C for 18 h.
10. After washing in 6X SSC with 0.1% sodium dodecyl sulfate at 23°C for 20 min and at 42°C for 20 min, perform audioradiography with Kodak XOMAT AR film (**Fig. 2**).

## 4. Notes

1. It is important to use uncharged slides. The use of highly charged slides will prevent easy capture of the cells. When performing immuno-LCM, one may lose some of the cells on the slides with the use of uncharged slides, so the procedures must be done very carefully.
2. It is important to use antibodies for immuno-LCM in as short a time as possible. Titration of the antibodies beforehand should optimize a condition. The use of RNasin may be helpful in preventing RNase contamination and degradation of the samples. Recent studies with antibodies combined for a direct or indirect immunostaining procedure should shorten the immunolabeling procedure.
3. We were able to obtain a strong signal from the GH3 pituitary cell lines using 1–10 cells to examine expression of specific hormones. We were able to capture up to 100 single cells in 30 min for multiple gene expression analyses.

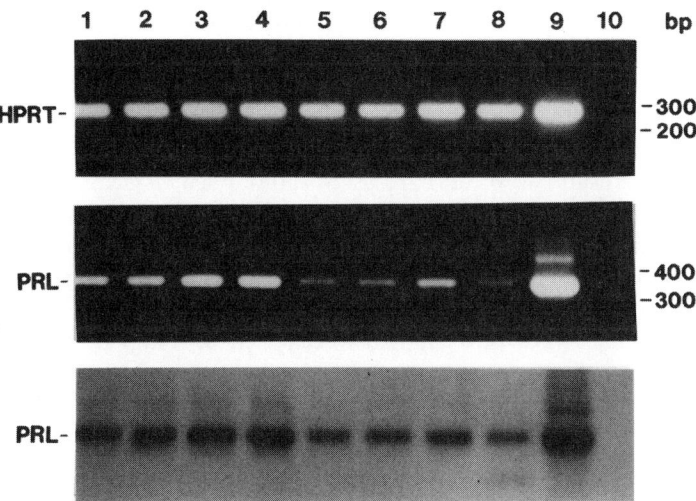

Fig. 2. RT-PCR analysis of GH3 cells (*n* = 10 cells/lane) after fixation in ethanol (lanes 1–4) or paraformaldehyde (lanes 5–8). Lane 9 represents a positive pituitary tissue control, lane 10 is negative control. (*Top*) HPRT is the housekeeping gene; (*middle*) the RT-PCR reaction product run on an ethidium bromide–stained gel is shown; (*bottom*) Southern hybridization with an internal probe for PRL is shown. These results show that fixation in ethanol is superior to fixation in paraformaldehyde for preservation of RNA quality and the recovery of RNA from cells.

4. TRIzol for RNA extraction or other buffers must be added to the collected cell as soon as possible to avoid degradation of nucleic acid and proteins.
5. In a study of sarcomas from fresh-frozen tissue sections, 1–10 cells were sufficient to detect translocation of genes for Ewing sarcoma, synovial sarcoma, alveolar rhabdomyosarcoma, and desmoplastic small round cell tumor using 40 cycles of RT-PCR (*10*). The bands were readily visualized after RT-PCR and gel electrophoresis. There was a three- to fivefold increase in the sensitivity of detecting the bands after Southern hybridization.
6. Various negative controls can be used for the standard RT-PCR reaction. Omission of the RT enzyme during the RT procedure is a rigorous control. Substituting water for the cDNA during the PCR can also be used as a control.

## References

1. Bonner, R. F., Emmert-Buck, M., Cole, K., Pohida, T., Chuaqui, R., Goldstein, S., and Liotta, L. A. (1997) Laser capture microdissection: molecular analysis of tissue. *Science* **278,** 1481–1483.
2. Burgess, J. K. and Hazelton, R. H. (2000) New developments in the analysis of gene expression. *Redox Rep.* **5,** 63–73.
3. Eltoum, I. A., Siegal, G. P., and Frost, A. R. (2002) Microdissection of histologic sections: past, present, and future. *Adv. Anat. Pathol.* **9,** 316–322.

4. Emmert-Buck, M. R., Bonner, R. F., Smith, P. D., et al. (1996) Laser capture microdissection. *Science* **274**, 998–1001.

5. Fend, F., Emmert-Buck, M. R., Chuaqui, R., Cole, K., Lee, J., Liotta, L. A., and Raffeld, M. (1999) Immuno-LCM: laser capture microdissection of immunostained frozen sections for mRNA analysis. *Am. J. Pathol.* **154**, 61–66.

6. Fend, F. and Raffeld, M. (2000) Laser capture microdissection in pathology. *J. Clin. Pathol.* **53**, 666–672.

7. Goldsworthy, S. M., Stockton, P. S., Trempus, C. S., Foley, J. F., and Maronpot, R. R. (1999) Effects of fixation on RNA extraction and amplification from laser capture microdissected tissue. *Mol. Carcinog.* **25**, 86–91.

8. Jin, L., Thompson, C. A., Qian, X., Kuecker, S. J., Kulig, E., and Lloyd, R. V. (1999) Analysis of anterior pituitary hormone mRNA expression in immunophenotypically characterized single cells after laser capture microdissection. *Lab. Invest.* **79**, 511, 512.

9. Jin, L., Tsumanuma, I., Ruebel, K. H., Bayliss, J. M., and Lloyd, R. V. (2001) Analysis of homogeneous populations of anterior pituitary folliculostellate cells by laser capture microdissection and reverse transcription-polymerase chain reaction. *Endocrinology* **142**, 1703–1709.

10. Jin, L., Majerus, J., Oliveira, A., Inwards, C. Y., Nascimento, A. G., Burgart, L. J., and Lloyd, R. V. (2003) Detection of fusion gene transcripts in fresh-frozen and formalin-fixed paraffin-embedded tissue sections of soft-tissue sarcomas after laser capture microdissection and RT-PCR. *Diagn. Mol. Pathol.* **12**, 224–230.

11. Rook, M. S., Delach, S. M., Deyneko, G., Worlock, A., and Wolfe, J. L. (2004) Whole genome amplification of DNA from laser capture-microdissected tissue for high-throughput single nucleotide polymorphism and short tandem repeat genotyping. *Am. J. Pathol.* **164**, 23–33.

12. Sansonno, D., Lauletta, G., and Dammacco, F. (2004) Detection and quantitation of HCV core protein in single hepatocytes by means of laser capture microdissection and enzyme-linked immunosorbent assay. *J. Viral Hepat.* **11**, 27–32.

13. Simone, N. L., Bonner, R. F., Gillespie, J. W., Emmert-Buck, M. R., and Liotta, L. A. (1998) Laser-capture microdissection: opening the microscopic frontier to molecular analysis. *Trends Genet.* **14**, 272–276.

14. Suarez-Quian, C. A., Goldstein, S. R., Pohida, T., et al. (1999) Laser capture microdissection of single cells from complex tissues. *Biotechniques* **26**, 328–335.

15. Todd, R., Lingen, M. W., and Kuo, W. P. (2002) Gene expression profiling using laser capture microdissection. *Expert Rev. Mol. Diagn.* **2**, 497–507.

# 3

## Fluorescence *In Situ* Hybridization on Single Cells

*Sex Determination and Chromosome Rearrangements*

### Paul N. Scriven and Caroline Mackie Ogilvie

### Summary

Fluorescence *in situ* hybridization (FISH) is the technique of choice for preimplantation genetic diagnosis (PGD) selection of female embryos in families with X-linked disease, for which there is no mutation-specific test. FISH with target-specific DNA probes is also the primary technique used for PGD detection of chromosome imbalance associated with Robertsonian translocations, reciprocal translocations, inversions, and other chromosome rearrangements, because the DNA probes, labeled with different fluorochromes or haptens, detect the copy number of their target loci. The methods described outline strategies for PGD for sex determination and chromosome rearrangements. These methods are assessment of reproductive risks, the selection of suitable probes for interphase FISH, spreading techniques for blastomere nuclei, and *in situ* hybridization and signal scoring using directly labeled and indirectly labeled probes.

**Key Words :** Fluorescence *in situ* hybridization; preimplantation genetic diagnosis; sex determination; translocations.

### 1. Introduction

The clinical application of preimplantation genetic diagnosis (PGD), after substantial investigations into the safety of biopsy procedures in mouse embryos, showed that the removal of one or two cells from an eight-cell human embryo did not significantly affect subsequent in vitro development to the blastocyst stage *(1)*. The first PGD pregnancies for couples who were at risk of transmitting X-linked disorders followed. The sex of the cell removed was determined using polymerase chain reaction of a Y-chromosome-specific repeat sequence with the transfer of embryos that had been genotyped as female *(2)*. However, failure of amplification of the Y-chromosome-specific sequence in a male embryo led to the first report of PGD misdiagnosis *(3)*. With the advent of fluorescence *in situ* hybridization (FISH) techniques and demonstration of the

From: *Methods in Molecular Medicine: Single Cell Diagnostics: Methods and Protocols*
Edited by: A. Thornhill © Humana Press Inc., Totowa, NJ

use of FISH as a robust technique for the preimplantation diagnosis of embryo sex *(4)*, FISH became the technique of choice for single cell diagnosis of sex chromosome status. FISH has the added advantage that the copy number of the target chromosome regions can be determined, and, therefore, the use of target-specific DNA probes labeled with different fluorochromes or haptens is also the primary technique used to detect chromosome imbalance associated with Robertsonian translocations, reciprocal translocations, inversions, and other chromosome rearrangements. Clinical application of PGD for chromosome rearrangements has employed chromosome paints and locus-specific probes using metaphase chromosomes (polar bodies *[5]*, blastomere nucleus conversion *[6,7]*) and interphase blastomere nuclei *(8–12)*. This chapter focuses on the use of FISH to determine sex and chromosome rearrangements in single embryonic cells.

## 2. Materials

1. Amine-coated slides (Genetix).
2. Diamond-tipped pen.
3. Cell lysis buffer (for spreading cells): 0.1% Tween-20 in 0.01 $M$ HCl, pH 2.0, at 20°C.
4. Phosphate-buffered saline (PBS), pH 7.0: 0.14 $M$ NaCl, 3 m$M$ KCl, 10 m$M$ $Na_2HPO_4$, 2 m$M$ $KH_2PO_4$.
5. Sterile distilled water.
6. Ethanol series: 70, 90, and 100%.
7. Low-power phase contrast microscope (x100).
8. Fluorochrome- or hapten-labeled *in situ* hybridization (ISH) DNA probes (Vysis, Oncor, CytoCell). Available fluorochromes/haptens and strategies for discriminating probes include the following:
   a. Probes labeled with Texas Red (TR).
   b. Probes labeled with fluorescein isothiocyanate (FITC).
   c. Probes labeled with SpectrumGreen (Vysis).
   d. Probes labeled with SpectrumOrange (Vysis).
   e. Vysis chromosome enumerator probes labeled with SpectrumAqua.
   f. Biotinylated probes detected with TR-avidin, FITC-avidin, or Cy-5 streptavidin (visualized using a FarRed filter).
   g. A mix of red and green probes to produce a yellow signal.
   h. A second round of hybridization *(13)*.
   i. A third color created using a SpectrumGold filter (*see* Note 1).
9. No. 0 cover slips (22 x 22 mm and 11 x 11 mm).
10. Denaturation hot block at 75°C (Hybrite or other brand calibrated to the exact temperature).
11. Rubber cement (Cow Gum; Cow Proofing, Slough, Berks, UK).
12. Incubator and hybridization chamber at 37°C.
13. 4X standard saline citrate (SSC)/Tween-20 wash solution: 0.05% Tween in 4X SSC, pH 7.0 (0.6 $M$ NaCl, 60 m$M$ $C_6H_5Na_3O$).

14. 0.4X SSC stringent wash solution: 0.4X SSC, pH 7.0, at 71°C.
15. Parafilm®.
16. 4N,6-Diamidino-2-phenylindole (DAPI)/Vectashield: 160 ng of DAPI in 1 mL of Vectashield mounting medium (Vector, Burlingame, CA).
17. Clear nail varnish.
18. Fluorescence microscope fitted with appropriate filters.
19. Imaging software (e.g., Isis, MetaSystems, Altlussheim, Germany; CytoVision, Applied Imaging, Newcastle upon Tyne, UK).
20. Where indirectly labeled probes are used, fluorescent avidin and fluorescent antidigoxygenin.

## 3. Methods

### 3.1. Reproductive Risk Assessment

#### 3.1.1. Sex-Linked Diseases

For female carriers of sex-linked diseases, assume 1:1 segregation of the sex chromosomes in both partners; fifty percent of male conceptions will have the defective X chromosome and, therefore, 25% of all conceptions will be affected by the disease. Live-born offspring are expected in the ratio of 1:1:1:<1 of normal females:carrier females:normal males:affected males. The risk of affected live-born male offspring will depend on the severity of the condition.

#### 3.1.2 Robertsonian Translocations

Robertsonian translocations are relatively common and the same translocation may be found in many different families. There are 10 heterozygous Robertsonian translocations; the most frequently encountered are der(13;14) and der(14;21). Segregation of translocations involving chromosomes 13 and 21 can result in zygotes with viable chromosome imbalance: translocation trisomy 13 (associated with Patau syndrome) and trisomy 21 (associated with Down syndrome). Reproductive risk estimates for these translocations are published and are based on collaborative studies; figures for the more common translocations are extrapolated to similar rare ones *(14,15)*.

#### 3.1.3. Reciprocal Translocations

Reciprocal translocations are carried by about 1 in 500 people and with one exception (the common t[11;22]; *see* Table 1) are effectively unique to the family in which they are found. The reproductive risk at conception and at term for the carrier of a reciprocal translocation is difficult to estimate because several possible modes of segregation occur with variable frequency, leading to many different theoretical permutations of each chromosome segment. The chromosome imbalance associated with the different permutations will determine how lethal the unbalanced translocation is *in utero*. For some translocations, very few, and in some cases none, of the abnormal meiotic products are compatible

**Table 1**
**Recommended Probe Sets for Common Chromosome Rearrangements**
(*see* **Note 1**)

| Rearrangement | Probe set |
|---|---|
| 45,Xn,der(13;14)(q10;q10) | Vysis LSI 13 (RB-1, 13q14, SpectrumGreen) |
| | CytoCell 13q subtelomere (D13S1825, 13q34, Texas Red) |
| | Vysis TelVysion 14q (D14S308, 14q32.3, SpectrumOrange) |
| 45,Xn,der(14;21)(q10;q10) | CytoCell 14q subtelomere (D14S1420, 14q32.3, Texas Red) |
| | Vysis LSI 21 ([D21S259,D21S341,D21S342], 21q22.13-q22.2, SpectrumOrange) |
| | CytoCell 21q subtelomere (D21S1575, 21q22.3, FITC) |
| 46,Xn,t(11;22)(q23.3;q11.2) | CytoCell 11q subtelomere (D11S4974, 11q25, Texas Red) |
| | Vysis LSI TUPLE 1 ([TUPLE 1,D22S553,D22S609, D22S942], 22q11.2, SpectrumOrange) |
| | Vysis LSI ARSA (ARSA, 22q13.3, SpectrumGreen) |

with life, and many may fail to implant. For other translocations, a pregnancy may be established that results in miscarriage, stillbirth, or the live birth of a child with profound mental and physical disabilities, which may result in death soon after birth. Translocation carriers may present with recurrent miscarriages, and some male carriers are infertile with spermatogenic arrest.

For most translocation carriers, the family pedigree is not extensive enough to allow a private risk assessment, and, therefore, reproductive risk assessment should be carried out as follows *(14,16 ,17)*:

1. Assess the mode of segregation that will give the smallest chromosome imbalance.
2. Use published empirical data on live-born children with single-segment imbalance for the regions involved.
3. Ascertain any relevant reproductive history in the family.
4. Review the literature.

### *3.2. Selection of Probe*

For PGD FISH assays, it is possible to combine directly labeled and indirectly labeled probes, and probes from different manufacturers. Probes for known polymorphic chromosome regions *(18,19)*, or those known to cross-hybridize significantly with other chromosomes *(20)*, should be avoided when possible. Ideally, to maximize the sensitivity of the test, every FISH assay should incorporate probes for each translocated chromosome or chromosome segment. However, this may not be possible if, e.g., chromosome rearrangement break points are distal to subtelomere probe regions, unique centromere region probes are not available, or suitable probes are not available either indirectly labeled with an appropriate hapten or directly labeled with a fluorochrome in an appropriate color. In addition, the use of, e.g., four probes instead of three for a reciprocal translocation is likely to increase the false positive error rate of the test.

### 3.2.1. Sex Determination

A probe set containing, as a minimum, three probes, specific for the centromere regions of the X and Y chromosomes and one autosome, is recommended for sex determination; the autosomal probe is used to establish ploidy. Such a probe set (usually alpha-satellite XY,18) should be chosen to have a very low expected polymorphism rate, and, therefore, precycle work-up should not be indicated *(21)*. Two scoring errors (failure to detect the Y chromosome signal in addition to detection of an extra X chromosome signal) are required to misdiagnose a male chromosome complement as a female complement; it is therefore acceptable to diagnose the sex of an embryo based on the result from a single biopsied cell.

### 3.2.2. Chromosome Rearrangements

The probe set should at least contain probes sufficient to detect all the expected forms of the rearrangement with chromosome imbalance. However, when suitable probes are not available, probe mixes that cannot detect some unbalanced forms of a rearrangement may be used provided that these unbalanced forms have been assessed to be nonviable in a recognizable pregnancy and/or to have a very low prevalence. Probe sets should be tested on cultured lymphocyte metaphases from both reproductive partners. At least 10 metaphase spreads should be examined to ensure that the probes are specific for the correct chromosomes; to assess chromosome polymorphism and signal cross-hybridization; and for a chromosome rearrangement carrier, to ensure that the probes hybridize as expected to the different segments of the rearrangement. In addition, at least 100 interphase nuclei from these preparations should be scored to assess signal specificity, brightness, and discreteness *(21)*. The interphase signal frequency data can be used to estimate the analytical performance of the single cell test (*see* **Note 2**).

#### 3.2.2.1. ROBERTSONIAN TRANSLOCATIONS

One probe for each of the two chromosomes involved in the translocation is required to detect all the theoretical unbalanced forms. However, if embryo diagnosis is to be based on the result from a single biopsied cell, then for

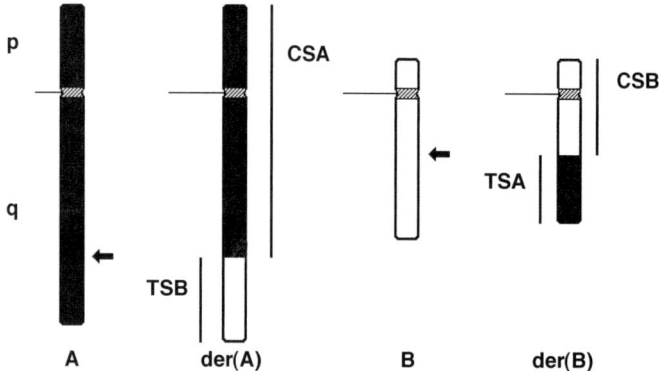

Fig. 1. Two-way reciprocal translocation between chromosomes A and B resulting in two derivative chromosomes der(A) and der(B). The portions exchanged are the translocated segments (TSA and TSB), and the rest of the chromosome that includes the centromere is the centric segment (CSA and CSB).

Robertsonian translocations involving chromosomes 13 and 21, two probes should be incorporated for each of these chromosomes, because they are associated with viable aneuploid pregnancies. Table 1 provides recommended probe sets for der(13;14) and der(14;21). If it is not possible to incorporate a second probe on the high-risk chromosome, then transfer should be based on a concordant result from two biopsied cells.

3.2.2.2. RECIPROCAL TRANSLOCATIONS

One probe for each of three of the four centric and translocated segments (*see* **Fig. 1**) is the minimum number of probes required to detect all the theoretical unbalanced forms of a reciprocal translocation *(22)*. The choice of probes should be based primarily on the predicted mode of segregation likely to result in live-born offspring with chromosome imbalance and the segregation mode considered likely to be most frequent. Adjacent-1 segregation, or alternate segregation following an odd number of crossovers in the interstitial segment (between the centromere and the break point), results in zygotes with monosomy for one translocated segment and trisomy for the other. Viable chromosome imbalance consistent with 3:1 segregation is usually associated with monosomy or trisomy for the smaller-derivative chromosome. Therefore, a centric segment probe for the smaller-derivative chromosome should be chosen if possible, in combination with a probe for each of the translocated segments. Very few reciprocal translocations result in viable offspring with imbalance owing to adjacent-2 segregation (monosomy and trisomy for the centric segments); however, where this is a possibility, probes for both centric segments

**Table 2**
**Reciprocal Translocation Segregation Modes and Chromosome Complements (A and B)**[a]

| Segregation mode | Chromosome complement | TSA | TSB | CSB |
|---|---|---|---|---|
| 4:0 Segregation | A,B | 1 | 1 | 1 |
| 3:1 Tertiary monosomy | A,der(A),B | 1 | 2 | 1 |
| 3:1 Interchange monosomy | A, B,B | 1 | 2 | 2 |
| Adjacent-2 crossover in A | A,der(A),der(A),B | 1 | 3 | 1 |
| Adjacent-1 | A,der(A),B,B | 1 | 3 | 2 |
| Adjacent-2 crossover in B | A,B,B,B | 1 | 3 | 3 |
| 3:1 Crossover in A | A,der(A),der(A),B,B | 1 | 4 | 2 |
| 3:1 Crossover in B | A,der(A),B,B,B | 1 | 4 | 3 |
| 3:1 Interchange monosomy | A,A,B | 2 | 1 | 1 |
| 3:1 Tertiary monosomy | A,B,der(B) | 2 | 1 | 2 |
| Adjacent-2 | A,A,der(A),B | 2 | 2 | 1 |
| Alternate | A,A,B,B | 2 | 2 | 2 |
| Alternate | A,der(A),B,der(B) | 2 | 2 | 2 |
| Adjacent-2 | A,B,B,der(B) | 2 | 2 | 3 |
| Anaphase II nondisjunction | A,A,der(A),der(A),B | 2 | 3 | 1 |
| 3:1 Tertiary trisomy | A,A,der(A),B,B | 2 | 3 | 2 |
| 3:1 Crossover in A | A,der(A),der(A),B,der(B) | 2 | 3 | 2 |
| 3:1 Interchange trisomy | A,der(A),B,B,der(B) | 2 | 3 | 3 |
| 3:1 Crossover in B | A,A,B,B,B | 2 | 3 | 3 |
| Anaphase II nondisjunction | A,B,B,B,der(B) | 2 | 3 | 4 |
| Adjacent-2 crossover in A | A,A,A,B | 3 | 1 | 1 |
| Adjacent-1 | A,A,B,der(B) | 3 | 1 | 2 |
| Adjacent-2 crossover in B | A,B,der(B),der(B) | 3 | 1 | 3 |
| Anaphase II nondisjunction | A,A,A,der(A),B | 3 | 2 | 1 |
| 3:1 Interchange trisomy | A,A,der(A),B,der(B) | 3 | 2 | 2 |
| 3:1 Crossover in A | A,A,A,B,B | 3 | 2 | 2 |
| 3:1 Tertiary trisomy | A,A,B,B,der(B) | 3 | 2 | 3 |
| 3:1 Crossover in B | A,der(A),B,der(B),der(B) | 3 | 2 | 3 |
| Anaphase II nondisjunction | A,B,B,der(B),der(B) | 3 | 2 | 4 |
| 4:0 Segregation | A,A,der(A),B,B,der(B) | 3 | 3 | 3 |
| 3:1 Crossover in A | A,A,A,B,der(B) | 4 | 1 | 2 |
| 3:1 Crossover in B | A,A,B,der(B),der(B) | 4 | 1 | 3 |

[a]The copy number of the translocated segment (TS) and the centric segments (CS) using the recommended one centric segment and both translocated segment probe scheme is given.

should be incorporated into the assay. Therefore, by incorporating informative probes for appropriate combinations of the translocated and centric segments, FISH assays can be designed to ensure that two scoring errors are required to diagnose a potentially viable chromosome imbalance as normal or balanced (*see* **Note 3** and **Table 2**). In most cases, diagnosis can therefore be based on the result from a single biopsied cell (*see* Note 4).

### 3.3. Spreading of Blastomeres

Methods of spreading and fixing single blastomeres include methanol/acetic acid *(23,24)* , Tween/HCl *(25)*, and a combination of Tween/HCl and methanol/acetic acid *(26)*.

Variations may include hypotonic treatment of cells prior to spreading and/or pepsin and paraformaldehyde treatment after fixation. The method should be appropriately validated for the laboratory *(21)*. In our laboratory, our experience has been almost exclusively with Tween/HCl, and this method is described in greater detail next.

### 3.3.1. Tween/HCl

This variation of the Tween/HCl method is technically simple and when standardized is highly reproducible in different laboratories. This method can be used to prepare single nuclei for the FISH diagnosis of sex determination and chromosome rearrangements:

1. Score a small circle (approx 5 mm) on the underside of an amine-coated slide and prelabel the slide with the case number, unique slide number, and biopsy date. Use a separate slide for each blastomere in numerical order, and label with the embryo number.
2. Place a small volume of lysis buffer (*see* **Note 5**) within the circle.
3. Transfer the blastomere into the lysis buffer; remove and add lysis buffer as necessary to lyse the cell.
4. Observe the nucleus to ensure that it remains within the circle and is not lost; if the cell is anucleate or multinucleated, biopsy another cell.
5. Leave the slide to air-dry at room temperature.

### 3.4. ISH and Signal Scoring

In most laboratories, methods of ISH employ codenaturation of the probe mix with the target DNA. The ISH method and signal-scoring criteria should be appropriately validated for the laboratory *(21)*. The variation of the codenaturation protocol described next is technically simple and can be used for sex determination and chromosome rearrangements:

1. Defrost probes, vortex, and centrifuge. Pipet volumes as required to make up the probe mixture (*see* **Note 5**).
2. Prewash fixed nuclei on slides using PBS for 5 min at room temperature.
3. Rinse twice in sterile distilled water.
4. Dehydrate with ethanol series (70, 90, and 100%) for 2 min each at room temperature and air-dry.

5. Record the position of the nucleus within the circle by visualizing with a phase contrast microscope.
6. Dehydrate with 100% ethanol for 2 min at room temperature and air–dry.
7. Apply 2 μL of probe mixture, and cover with a 9 × 9 mm no. 0 cover slip.
8. Seal the edges of the cover slip with rubber solution.
9. Codenature the slides on a hot block (e.g., Hybaid Omnislide or Vysis Hybrite) at 75°C for 5 min, and then hybridize the slides overnight (16–20 h) in a humidified chamber at 37°C. Probe mixes that consist entirely of centromere probes (i.e., for sex-linked cases) will give a satisfactory result after 60 min of hybridization.
10. Carefully remove the rubber solution from each slide and rinse off the cover slip using 4X SSC/0.05% Tween-20 at room temperature.
11. Wash the slides in a 0.4X SSC stringent wash at 71°C for 5 min.
12. Wash the slides in 4X SSC/0.05% Tween-20 at room temperature for 2 min.
13. If the probe mix contains indirectly labeled probe(s), drain the slides of excess liquid and apply 20 μL of fluorescently conjugated antibody under a 20 × 20 mm square of Parafilm. Incubate in a humidified chamber at 37°C for 15 min. Remove the Parafilm and wash once in 4X SSC/0.05% Tween-20 at room temperature for 2 min.
14. Wash twice for 2 min in PBS at room temperature and drain the slides.
15. Apply 6 μL of DAPI to a 22 × 22 mm no. 0 cover slip and invert the slide over the cover slip.
16. Blot and seal the edges of the cover slip with clear nail varnish.
17. Analyze using a fluorescence microscope suitably equipped with the appropriate filter sets for the probes being used.

### 3.4.1. Signal Scoring

Score signals by direct visualization using a fluorescence microscope and single band-pass filters for each fluorochrome in the assay (*see* **Note 6**). Use imaging software to capture an image of the nucleus for confirmation of the visual diagnosis, and for image archiving as part of the laboratory quality assurance plan *(21)*.

### 3.4.2. Intraassay Controls

The value of positive and negative control slides for FISH is debatable. A negative control slide with metaphase spreads can be used to confirm that the correct probes were applied. Owing to the large number of possible combinations of monosomy and trisomy for the different chromosome segments, a suitable positive control slide for most chromosome rearrangements is generally not available; however, some groups use triploid material, which has trisomy for all the segments involved in the chromosome rearrangement. We take the view that for FISH control slides are of very limited value and that owing to the possibility of contamination it is undesirable to process control slides containing thousands of nuclei alongside slides with only a single nucleus. To ensure that the correct probes are used, two observers can check the case details and probes independently before they are applied; the different embryos, to some extent, provide positive and negative controls within each cohort.

## 4. Notes

1. An elegant method to generate a third pseudocolor exploits the observation that Vysis SpectrumOrange probes can be visualized using a Texas Red filter and a SpectrumGold filter although CytoCell Texas Red–labeled probes can only be seen using the Texas Red filter (personal communication, 2003 Julie Oliver, genetic technologist, Genetics & IVF Institute, Fairfax, VA).

2. Analytical performance of the single cell test is assessed based on the lymphocyte work-up by estimating the false positive error rate (multiplying the probability of a normal signal pattern for each probe) and the false negative error (1–the sum of the probabilities of a normal signal pattern for each of the unbalanced forms of the translocation), and estimating the negative predictive value (the likelihood that a negative test result is a normal or balanced translocation chromosome complement) and the positive predictive value (the likelihood that a positive test result is an unbalanced translocation chromosome complement). Where there are two informative probes for potentially viable imbalance, an estimated negative predictive value of at least 0.95 and a positive predictive value of at least 0.85 are considered acceptable in order to offer a clinical PGD cycle using single cell biopsy.

3. There must be a balance of priorities between establishing a successful pregnancy and minimizing the risk of an affected outcome; an overcautious PGD strategy may result in marginal reduction in the risk of misdiagnosis at the expense of live births. Single cell biopsy and judicious selection of three FISH probes (one centric segment and both translocated segments; *see* Table 2) is recommended as the optimum strategy for most translocations *(22)*. However, incorporation of a fourth probe and/or transfer based on a concordant two-cell result will be indicated in some cases.

4. It is prudent to consider transfer based on a concordant two-cell result for which imbalance for small-derivative chromosomes is known or is likely to be viable. A small-derivative chromosome may be lost from a balanced chromosome complement or 3:1 segregation, resulting in tertiary trisomy for the small-derivative chromosome.

5. Two batches of lysis buffer are available for each cycle: a batch that was made fresh the day before and stored in 1-mL aliquots at –20°C, which is used in the first instance; and a backup batch that was used and found to be satisfactory from the previous cycle. All probe vials are tested before clinical application, to confirm that they contain the correct chromosome-specific probe labeled with the correct fluorochrome or hapten, and to assess that signal specificity, brightness, and discreteness are within acceptable parameters.

6. A general guideline should lead to scoring of a single signal where two closely spaced signals are less than one domain apart; however, judgment based on experience needs to be exercised to interpret signals of varying size, intensity, and separation.

## References

1. Hardy, K., Martin, K. L., Leese, H. J., Winston, R. M., and Handyside, A. H. (1990) Human preimplantation development in vitro is not adversely affected by biopsy at the 8-cell stage. *Hum. Reprod.* **5**, 708–714.
2. Handyside, A. H., Kontogianni, E. H., Hardy, K., and Winston, R. M. (1990) Pregnancies from biopsied human preimplantation embryos sexed by Y-specific DNA amplification. *Nature* **344**, 768–770.
3. Hardy, K. and Handyside, A. H.(1992) Biopsy of cleavage stage human embryos and diagnosis of single gene defects by DNA amplification. *Arch. Pathol. Lab. Med.* **116**, 388–392
4. Griffin, D. K., Handyside, A. H., Harper, J. C., et al. (1994) Clinical experience with preimplantation diagnosis of sex by dual fluorescent *in situ* hybridization. *J. Assist. Reprod. Genet.* **11**, 132–143.
5. Munne, S., Morrison, L., Fung, J., et al. (1998) Spontaneous abortions are reduced after preconception diagnosis of translocations. *J. Assist. Reprod. Genet.* **15**, 290–296.
6. Willadsen, S., Levron, J., Munne, S., Schimmel, T., Marquez, C., Scott, R., and Cohen, J. (1999) Rapid visualization of metaphase chromosomes in single human blastomeres after fusion with in-vitro matured bovine eggs. *Hum. Reprod.* **14**, 470–475.
7. Verlinsky, Y., Cieslak, J., Evsikov, S., Galat, V., and Kuliev, A. (2002) Nuclear transfer for full karyotyping and preimplantation diagnosis for translocations. *Reprod. Biomed. Online.* **5**, 300–305.
8. Pierce K. E., Fitzgerald, L. M., Seibel, M. M., and Zilberstein, M. (1998) Preimplantation genetic diagnosis of chromosome balance in embryos from a patient with a balanced reciprocal translocation. Mol. Hum. Reprod. **4**, 167–172.
9. Munne, S., Sandalinas, M., Escudero, T., Fung, J., Gianaroli, L., and Cohen, J. (2000) Outcome of preimplantation genetic diagnosis of translocations. *Fertil. Steril.* **73**, 1209–1218.
10. Ogilvie, C. M., Braude, P., and Scriven, P. N. (2001) Successful pregnancy outcomes after preimplantation genetic diagnosis (PGD) for carriers of chromosome translocations. *Hum. Fertil.* **4**, 168–171.
11. Fridstrom, M., Ahrlund-Richter, L., Iwarsson, E., et al. (2001) Clinical outcome of treatment cycles using preimplantation genetic diagnosis for structural chromosomal abnormalities. *Prenat. Diagn.* **21**, 781–787.
12. ESHRE PGD Consortium Steering Committee. (2002) ESHRE Preimplantation Genetic Diagnosis Consortium: data collection III (May 2001). *Hum. Reprod.* **17**, 233–246.
13. Ogur, G., Van Assche, E., and Liebaers, I. (2002) Preclinical work-up of preimplantation genetic diagnosis for chromosomal translocation carriers. In: Macek Sr M, Bianchi DW, Cuckle H (eds). *Early Prenatal Diagnosis, Fetal Cells and DNA in the mother. Present State and Perspectives.* Prague: Charles University in Prague, The Karolinum Press, 236–253.
14. Gardner, R. J. M. and Sutherland, G. R. (2004) *Chromosome Abnormalities and Genetic Counseling.* 3rd Ed. New York: Oxford University Press.

15. Scriven, P. N., Flinter, F. A., Braude, P. R., and Ogilvie, C. M. (2001) Robertsonian translocations—reproductive risks and indications for preimplantation genetic diagnosis. *Hum. Reprod.* **16,** 2267–2273.
16. Midro, A. T., Stengel-Rutkowski, S., and Stene, J. (1992) Experiences with risk estimates for carriers of chromosomal reciprocal translocations. *Clin. Genet.* **41,** 113–122.
17. Scriven, P. N. (2003) Preimplantation Genetic Diagnosis for Carriers of Reciprocal Translocations. *J. Assoc. Genet. Technol.* **29,** 49–59.
18. Hsu, L. Y., Benn, P.A., Tannenbaum, H. L., Perlis, T. E., Carlson, A. D. (1987) Chromosomal polymorphisms of 1, 9, 16, and Y in 4 major ethnic groups: a large prenatal study. *Am. J. Med. Genet.* **26,** 95–101.
19. Shim, S. H., Pan, A., Huang, X. L., Tonk, V. S., Varma, S. K., Milunsky, J. M., .and Wyandt, H. E. (2003) FISH Variants with D15Z1. *J. Assoc. Genet. Technol.* **29,** 146–151.
20. Knight, S. J. and Flint, J. (2000) Perfect endings: a review of subtelomeric probes and their use in clinical diagnosis. *J. Med. Genet.* **37,** 401–409.
21. Thornhill A. R., deDie-Smulders C. E., Geraedts J. P., et al. (2005). ESHRE PGD Consortium 'Best practice guidelines for clinical preimplantation genetic diagnosis (PGD) and preimplantation genetic screening (PGS)'. *Hum. Reprod.* **20,** 35–48.
22. Scriven, P. N., Handyside, A. H., and Ogilvie, C. M. (1998) Chromosome translocations: segregation modes and strategies for preimplantation genetic diagnosis. *Prenat. Diagn.* **18,** 1437–1449.
23. Tarkowski, A. K. (1966) An air drying method for chromosome preparations from mouse eggs. *Cytogenetics.* **5,** 394–400.
24. Munné, S., Lee, A., Rosenwaks, Z., Grifo, J. and Cohen, J. (1993) Diagnosis of major chromosome aneuploidies in human preimplantation embryos. *Hum. Reprod.* **8,** 2185–2192.
25. Harper, J. C., Coonen, E., Ramaekers, F. C. S., Delhanty, J. D. A., Handyside, A. H., Winston, R. M. L., and Hopman, A. H. N. (1994) Identification of the sex of human preimplantation embryos in two hours using an improved spreading method and fluorescent in situ hybridisation using directly labelled probes, *Hum. Reprod.* **9,** 721–724.
26. Dozortsev D. I. and McGinnis K. T. (2001) An improved fixation technique for fluorescence in situ hybridization for preimplantation genetic diagnosis. *Fertil. Steril.* **76,**186–8.

# 4

## Single Cell Polymerase Chain Reaction for Preimplantation Genetic Diagnosis

*Methods, Strategies, and Limitations*

### Karen Sermon and Martine De Rycke

### Summary

We provide an overview of the methodology involved in single cell polymerase chain reaction (PCR), especially for single lymphocytes or cultured lymphoblasts and blastomeres. The first step toward single cell PCR is isolation of single cells; the protocols given can be carried out using basic instruments such as a stereomicroscope. We also describe the alkaline lysis method for cell lysis as well as the design and execution of single cell PCR, either in simplex or in multiplex. Finally, we briefly discuss the different methods for analyzing PCR products and obtaining an accurate diagnosis. Special attention is given to measures that avoid contamination with extraneous DNA and reduce allele dropout.

**Key Words:** Single cell PCR; preimplantation genetic diagnosis; fluorescent PCR; multiplex PCR; contamination; allele dropout.

## 1. Introduction

Preimplantation genetic diagnosis (PGD) is an early form of prenatal diagnosis in which embryos obtained in vitro are analyzed and assigned a diagnosis before they are transferred to the mother, thus avoiding the need for termination of pregnancy after a negative result following a prenatal diagnosis. The diagnosis can be obtained from the first and the second polar bodies of oocytes biopsied soon after retrieval of the oocytes, or on single embryonic cells (blastomeres) obtained after biopsy on d 3. PGD became possible only after polymerse chain reaction (PCR) could be refined to a point at which the DNA of a single cell could be sufficiently amplified to allow analysis. The first article describing PGD appeared in the early 1990s and described the sexing of embryos through the amplification of Y-specific repetitive sequences *(1)*. Since then, fluorescence *in situ* hybridization has become the method of choice for

From: *Methods in Molecular Medicine: Single Cell Diagnostics: Methods and Protocols*
Edited by: A. Thornhill © Humana Press Inc., Totowa, NJ

sexing embryos, but PCR has been further refined to make diagnosis possible for a myriad of monogenic diseases at the single cell level *(2)*.

The following steps can be distinguished in single cell PCR:

1. Obtaining a single cell from a group (sample, culture, embryo) and transferring this cell to a PCR tube.
2. Lysis of the cell: Classic extraction of DNA with phenol/chloroform or columns is not possible, so the cell is lysed in the tube and the PCR components are added (single-tube method).
3. Single cell PCR: The PCR components are added to the lysed cell, so changes in concentrations of components and pH have to be taken into account. In addition, in multiplex PCR, an ideal balance between the amplifications of different amplicons must be found. To compensate for possible interactions between primers, or for the fact that some primer pairs are less efficient than others or have widely varying annealing temperatures, a second PCR round can be introduced.
4. Analysis of the PCR products: This can be carried out on agarose gels, but more efficient fluorescent PCR and analysis on an automated DNA analyzer is recommended. The detection of some mutations may require post-PCR assays (restriction or minisequencing). Finally, it must be possible to perform the whole procedure within a short time frame (i.e., 2 d if the embryos were biopsied at the cleavage stage). Indeed, with the currently available embryo culture methods, embryos cannot be kept in vitro much longer than d 5. Cryopreservation of embryos can circumvent this problem, but the transfer of fresh embryos still has a greater chance of implantation and ongoing pregnancies.

Centers wishing to start PGD for monogenic diseases and looking for consensus guidelines outlining the setup of the clinic, laboratory, and other technical aspects of PGD are referred to the ESHRE PGD Consortium guidelines *(3)*.

## 2. Materials

### 2.1. Collection of Cells: Lymphocytes or Lymphoblasts and Research Blastomeres

#### 2.1.1. Equipment

1. Dedicated laminar flow with stereomicroscope (*see* **Note 1**).
2. Centrifuge.
3. Filtered tips: good-quality tips that allow maximum recovery (e.g., Axygen MAXYMum Recovery filter tips, presterilized) are recommended.
4. Gloves: must be close fitting (*see* **Note 1**).
5. Capillaries: microcaps (30 µL) (Drummond).
6. Hand-drawn glass Pasteur pipets or Stripper Tips (75µm id; ref. no. MXL3; Mid-Atlantic Diagnostics).
7. PCR tubes (0.2 mL): Because the transfer of blastomeres can be followed through the stereomicroscope, the plastic must be translucent. We currently use Axygen

Thin Wall PCR tubes. Colored tubes are useful for distinguishing different reactions during two-round PCRs.

## 2.1.2. Solutions and Products

When possible, solutions must be autoclaved and aliquoted. An autoclave dedicated for single cell PCR must be used.

1. Lymphoprep® (ref. no. 1114545; Axis-Shield, Rodeløkka, Norway).
2. Phosphate-buffered saline (PBS).
3. In-house-made $Ca^{2+}$- and $Mg^{2+}$-free medium: 14 m$M$ NaCl, 0.2 m$M$ KCl, 0.04 m$M$ $NaH_2PO_4 \cdot 2H_2O$, 5.5 m$M$ glucose, 1.2 m$M$ $NaHCO_3$, 0.02 m$M$ EDTA, 0.01% (w/v) Phenol Red. For 100 mL, make up 0.8 g of NaCl, 0.020 g of KCl, 0.005 g of $NaH_2PO_4$, 0.100 g of glucose, 0.100 g of EDTA, 0.100 g of $NaHCO_3$, 0.010 g of Phenol Red, and distilled water to 100 mL. Autoclave and dispense into 1-mL aliquots for single use. This solution is commercially available from a number of vendors.
4. Mineral oil (M8410; Sigma, St. Louis, MO).
5. Alkaline lysis buffer (ALB): 50 m$M$ dithiothreitol (DTT), 200 m$M$ KOH or NaOH. To make this solution, use 0.0077 g of DTT, 1 $M$ (200 µL) KOH or NaOH, and 800 µL of MilliQ water. A series of tubes with the DTT preweighed can be prepared and stored for a long time at 4°C, and the appropriate amount of KOH/NaOH and water added just prior to use. ALB should be kept at 4°C and not be used for longer than 1 wk.
6. Acidic Tyrode's solution: 0.8 g of NaCl, 0.02 g of KCl, 0.024 g of $CaCl_2 \cdot 2H_2O$, 0.01 g of $MgCl_2 \cdot 6H_2O$, 0.1 g of glucose. Bring to pH 2.4, autoclave, and aliquot. Alternatively, acidic Tyrode's solution is commercially available from a number of vendors.
7. Deaza-dGTP, (General Electric Biotechnologies).

## 2.2. Polymerase Chain Reaction

### 2.2.1. Equipment

It should be emphasized that only PCR machines of good quality are suitable for single cell PCR. Because single cell PCR is extremely sensitive to suboptimal experimental conditions, it is essential to invest in PCR equipment with a high level of reproducibility. In our experience, Applied Biosystems and Eppendorf machines meet these high standards. The Mastercycler gradient PCR from Eppendorf is very useful for optimization of annealing temperature.

### 2.2.2. Solutions and Products

1. DNA polymerase: AmpliTaq DNA polymerase with GeneAmp 10XPCR buffer (ref. no. N808-0153; Applied Biosystems).
2. dNTPs: DNA Polymerisation Mix (ref. no. 27-2094-01; General Electric Biotechnology).

3. Primers (Eurogentec, Seraing, Belgium).
4. 10X PCR buffer without KCl: 100 m$M$ Tris-HCl, pH 8.3; 20 m$M$ MgCl$_2$; 0.1% gelatin. To make up, use 5 mL of 2 $M$ Tris-HCl, pH 8.3; 2 mL of 1 $M$ MgCl$_2$; and 0.1 g of 0.1% (w/v) gelatin. Add MilliQ water to 100 mL, autoclave, and aliquot.
5. 10X Neutralization buffer: 900 m$M$ Tris-HCl, pH 8.3; 300 m$M$ KCl; 200 m$M$ HCl. To make up, use 450 μL of 2 $M$ Tris–HCl, pH 8.3; 150 μL of 2 $M$ KCl; 200 μL of 1 $M$ HCl; and 200 μL of H$_2$O.
6. 10X Tricine: 500 m$M$ tricine, pH 4.9 (T-6272; Sigma). To make up, dissolve 3.58 g in 100 mL of H$_2$O. Bring to pH 4.9 with NaOH before autoclaving and aliquoting.
7. Special kits (polymerase + buffer):
   a. Expand High Fidelity PCR System 2 X 250U (ref. no. 1 732 650; Roche).
   b. Expand Long Template PCR system 2 X 360U (ref. no. 1 681 842; Roche).
8. Dimethylsulfoxide (DMSO) (D8779; Sigma).

## 3. Methods

### 3.1. Collection of Lymphocytes and Lymphoblasts

Lymphocytes can be separated from a fresh blood sample (10 cc on heparin) using Lymphoprep. Epstein-Barr virus–transformed lymphoblasts can also be used as a model to develop single cell PCRs. The lymphoblasts are cultured as previously described (*4*).

Single lymphocytes or lymphoblasts are collected as follows:

1. For the lymphoblasts, collect a few colonies, transfer to a 1.5-mL PCR tube, and allow to settle to the bottom.
2. Remove the supernatant in order to discard the dead cells floating free in the medium and to collect mainly the living cells present in the colonies.
3. Wash the cells three times by centrifuging for 5 min at 4000 rpm, discard the supernatant, resuspend the pellet in 500 μL of PBS, and perform a final resuspension in 50 μL of PBS. The lymphocytes and lymphoblasts suspended in PBS can be kept at 4°C for a maximum of 3 d, but it must be taken into account that the quality of the cells decreases each day (*see* **Note 2**).
4. Prepare a Petri dish with Ca$^{2+}$- and Mg$^{2+}$-free medium droplets supplemented with bovine serum albumin (BSA) (15 mg/mL) to prevent cell adherence to the plastic of the dish. At the top, make two 5-μL droplets. Then make rows of three smaller, 2μL droplets. Cover the drops with 5 mL of mineral oil.
5. Using a hand-drawn microcapillary controlled with a mouthpiece and under stereomicroscopic control, transfer a large group of cells to the large droplets. Using a freshly made microcapillary, transfer a small group of 10–20 cells to the first of the three 2-μL droplets. From there, one cell can be picked up, washed twice in the second and third droplet, and then transferred without microscopic observation to 200-μL PCR tubes containing 2.5 μL of ALB. Rinse the microcapillary between cells and before taking a blank, which is a small sample taken from the last washing droplet. Take one blank for each cell collected. Keep the PCR tubes with the cells

and the blanks on ice during the whole procedure and freeze at –80°C until further processing.

### 3.2. Collection of Blastomeres

### 3.2.1. Dissociation of Whole Embryos into Single Blastomeres

During the preliminary setup of the PGD assay, it is sometimes required that PCR be applied to single blastomeres. Cleavage-stage embryos that have not undergone compaction yet and are donated for research are the most suitable. Blastocysts are not suitable for single cell analysis owing to problems with disaggregation of individual cells.

A Petri dish is prepared with 5-μL droplets of acidic Tyrode's solution and $Ca^{2+}$- and $Mg^{2+}$-free medium supplemented with 4 mg/mL of BSA and covered with mineral oil. This dish must be at 37°C for optimal activity of the acidic Tyrode's. Using a drawn Pasteur pipet and a mouth pipet, the embryo is transferred under stereomicroscopic control to the first drop of acidic Tyrode's, and then to the next drop. Care must be taken to transfer as little as possible of the embryo culture medium to the acidic Tyrode's to preserve the correct pH. Usually, the zona pellucida (ZP) disappears within a few seconds, so the embryo must stay under visual control. Removal of the ZP can be aided by pipetting the embryo up and down in the drawn Pasteur pipet. After washing the embryo two times in $Ca^{2+}$- and $Mg^{2+}$-free medium, it can be left in this medium at 37°C for the time necessary for dissociation. This can be quite variable and can take between seconds and minutes. Again, the dissociation of the embryo can be aided by pipetting it up and down in the Pasteur pipet.

### 3.2.2. Transfer of a Single Blastomere to PCR Tube

Once the embryo is dissociated, blastomeres with a clearly visible nucleus can be selected, to be transferred to a 200-μL PCR tube. The cell is washed three times in 5-μL droplets of $Ca^{2+}$- and $Mg^{2+}$-free medium under mineral oil supplemented with 5 mg/mL of BSA at room temperature to avoid adherence of the cell to the Petri dish. During the clinical PGD, single blastomeres are obtained after biopsy (5). The blastomere can be transferred to the PCR tube blindly, or visualized under stereomicroscope control as follows:

1. Take the blastomere into a drawn Pasteur pipet or a Stripper Tip, taking care to leave the cell as close as possible to the end of the pipet or tip.
2. Lay the PCR tube containing 2.5 μL of ALB on its side with the opening facing right. Fix the focus of the stereomicroscope on the top meniscus of the ALB droplet.
3. Gently introduce the Pasteur pipet sideways into the PCR tube, taking care to direct the tip to the top wall of the tube so that it stays in focus. Once the tip

reaches the ALB droplet, gently blow the blastomere out of the pipet into the lysis buffer. If the lysis buffer works properly, the blastomere should slowly swell and then disintegrate.

Keep the PCR tubes on ice during the whole procedure (except when the blastomere is introduced under the stereomicroscope), and freeze at –80°C as soon as possible. In our experience, the best amplification results are achieved when the cells are frozen for at least 30 min and for a maximum of 1 wk (*see* **Note 2**).

### *3.3. PCR of Single Cells*

#### *3.3.1. Cell Lysis*

When taken from the freezer, the cells are kept on ice and then transferred to a PCR machine. They are incubated at 65°C for 10 min and then directly brought to 4°C again (either in the PCR machine or on ice).

#### *3.3.2. Choice of Primers*

The choice of the correct primers is essential for a successful single cell PCR, especially if duplex or multiplex PCRs are applied. The sequence of the primers can be extracted from the literature or it can be designed in-house, possibly with the help of software programs (e.g., Primer 3, available free at http://frodo.wi.mit.edu/cgi-bin/primer3/primer3_www.cgi, or Primer Express from Applied Biosystems), using sequence information available on Ensemble, The Wellcome Trust Sanger Institute (http://www.ensembl.org/) and the National Center for Biotechnology Information nucleotide database (http://www.ncbi.nlm.nih.gov/). The first alternative is often used for complex DNA regions (e.g., large and complex genes with pseudogenes), while the second is suitable for known DNA sequences, such as gene exons and microsatellites. Several software packages are available to test the primers *in silico* (e.g., Primer 3 and Primer Express) as a pair to amplify one sequence, but also when mixed with other primers. It is strongly recommended to check, again, *in silico* the specificity of each primer by running its DNA sequence in the BLAST program (http://www.ncbi.nlm.nih.gov/BLAST/) and to use only primers with a unique match to the region of interest.

For duplex or multiplex PCR, the following points must be taken into account:

1. The different primers must not interact; for example, no complementary sequences must be present. This can be checked with software programs such as Primer 3.
2. The length of the amplicons must be taken into account, especially if an automated DNA analyzer that can read only one fluorochrome (e.g., AlfExpress) is used. Many DNA analyzers can read up to five fluorochromes, so care must be taken not

to label two primer pairs generating amplicons of the same length with the same fluorochrome. Generally, small amplicons replicate more efficiently and will reduce the level of allele dropout (ADO) (*see* **Note 3** and **ref.** *6*). We usually amplify fragments in a range of 80–350 bp.

3. The annealing temperature of the primers must be as similar as possible. A compromise that most probably will lie closer to the lowest annealing temperature can usually be found. The GC content of the primers must be similar and range between 40 and 60%.

4. The efficiency of amplification, mirrored in the number of cycles needed, can vary widely between primer sets and is usually more efficient for exonic sequences than for repetitive sequences. This efficiency can also vary between batches of the same primer, depending on the primer synthesis and labeling efficiency (*see* **Note 2**).

5. GC-rich DNA sequences will require the use of special additives to the PCR. These additives facilitate the denaturation of the amplicons by lowering the denaturation temperature (e.g., DMSO), or by using 7-deaza-2N-deoxyguanosine (deaza-dGTP,), a dGTP analog that incorporates into the DNA instead of the dGTP and that binds the two DNA strands less tightly together.

### 3.3.3. Choice of Amplification Systems

By amplification system we mean the combination of DNA polymerase enzyme and appropriate buffer. At our laboratory, we have used three systems with good results:

1. Taq polymerase: We routinely use AmpliTaq polymerase, whereas the AmpliTaq Gold system does not work for us. The advantage of this system is that the buffer is quite simple and the composition is known, so that an adapted in-house reaction buffer containing no KCl can be made. This is necessary because the ALB buffer and the neutralization buffer both contain KCl and KCl at high concentrations will inhibit the PCR. Our experience is that AmpliTaq polymerase works for simple sequences (not repetitive or CG-rich) in simplex PCRs but is less well adapted for duplex or multiplex PCRs.

2. Expand High Fidelity polymerase (EHF) (Roche): This is a mixture of two DNA polymerases (Taq polymerase and Tgo polymerase) and has a better proofreading activity than AmpliTaq. We routinely use EHF if the AmpliTaq does not give expected results. It performs very well for microsatellites and in duplex and multiplex PCRs. EHF comes with two buffers (one with and one without $MgCl_2$), and these buffers can easily be combined with ALB with KOH or NaOH. DMSO can easily be added to the system.

3. Expand Long Template polymerase (ELT) (Roche): This contains the two same DNA polymerases as EHF, but in different ratios. It is conceived for the amplification of long templates but is very useful for the amplification of GC-rich templates. It comes with three different buffers that can be used according to the length of the amplicon. DMSO and deaza-dGTP can easily be combined in this system. ELT often is not suitable for shorter amplicons that work well with EHF and is reserved

for CG-rich templates such as triplet repeats (e.g., the CGG triplet in Fragile-X syndrome).

### 3.3.4. Choice of dNTPs

Most companies offer dNTPs as four separate tubes or premixed. The separate tubes are useful when deaza-dGTP needs to be added. When deaza-dGTP is used, we routinely add it in a 1:3 ratio with dGTP.

### 3.3.5. Different Steps in Development of a Single Cell PCR

Newly synthesized primers are first tested on genomic DNA in large amounts such as 100 ng. The exact annealing temperature can be established using a PCR machine with a temperature gradient function (e.g., Eppendorf Mastercycler Gradient), by testing a temperature range 5°C below and 5°C above the calculated annealing temperature of the primers. The primers must be tested on normal samples and samples from the patients (carriers and homozygous affected for recessive diseases). If a duplex or multiplex PCR is to be developed, the different primers can be tested together at this point, by choosing a common annealing temperature that is as low as possible or, again, by performing an annealing temperature gradient. It may become obvious here that the different primer sets cannot be used together. A possibility then is to perform 10 PCR cycles with all the primers together and then split up the different reactions by either keeping together primers that perform well in the same reaction or splitting all primer sets up in different reactions.

The next step is the downscaling to 100 pg of genomic DNA (this is equivalent to about 16 single cells). At this point, more PCR cycles will have to be run, but adjusting the annealing temperature, the choice of DNA polymerase, and the use of additives such as DMSO may be necessary. In a multiplex, the concentration of the different primer sets must be adjusted so that the amplification yield for each amplicon is similar. The next step is to go to testing at the single cell level. For some primers, there may be a significant difference between the efficiency at 100 pg and at the single cell level. This can usually be overcome by increasing again the number of cycles, but occasionally it will be necessary to order new primer sets, either the same sequences, because the batch quality was inferior, or completely newly designed primers. An alternative strategy to overcome inefficient amplification is the use of heminested or nested primers. Such primers raise the specificity; however, the overall PCR becomes more complex and prone to errors.

### 3.3.6. Different Steps in Single Cell PCR

1. Before lysis of the cells, prepare reaction mix sufficient for all the samples. This reaction mix typically contains 200 $\mu M$ dNTPs that are usually kept in aliquots at

a 10X concentration, 1X reaction buffer (homemade without KCl if AmpliTaq is used, provided by the manufacturer if EHF is used), 1X neutralization buffer if AmpliTaq is used or 1X Tricine buffer if EHF or ELT is used, 20 $\mu M$ primers, additives such as DMSO (between 0 and 15 %), and autoclaved MilliQ water. The DNA polymerase is added later. Keep the mix at 4°C until further use. It is recommended that the mix be used as soon as possible, because primers can start to anneal in the tube even at low temperatures. Prepare the reaction mix in a pre-PCR laminar flow to minimize the risk of contamination (*see* **Note 1**).

2. In a different laboratory, have a DNA dilution of the appropriate sample (e.g., carrier of the mutation to be amplified) diluted to 100 pg/$\mu$L. Only single cells and 100 pg/$\mu$L DNA dilutions can be brought into the pre-PCR laminar flow.
3. Lyse the cells by incubating for 10 min at 65°C in a PCR machine and immediately put on ice after lysis.
4. In the pre-PCR flow, carefully open all the tubes containing the single cells. Add 1 $\mu$L of the 100 pg/$\mu$L DNA dilution to the positive control tube. Add the appropriate amount of DNA polymerase to the cooled reaction mix and briefly vortex. Then add the complete reaction mix to the lysed cells. We usually pick up cells in 2.5 $\mu$L of ALB, and add reaction mix to a total volume of 25 $\mu$L.
5. Immediately start the PCR reactions, preferably in a heating block that has already been preheated to the required denaturing temperature (95°C).
6. If a second round of PCR is to be performed, dispense the reaction mixes with the DNA polymerase into fresh PCR tubes (colored tubes can be useful at this point). To avoid keeping the reaction mixes at 4°C for a longer time, they can be prepared while the first PCR round is running. Add 2–4 $\mu$L of the first-round PCR product to the mixes without delay. Immediately start the PCR.

### 3.3.7. Analysis of PCR Fragments

It is beyond the scope of this chapter to illustrate and discuss all the possibilities of analysis of PCR fragments. Nevertheless, the most important methods currently used in PGD are briefly discussed next.

1. Detection on agarose or polyacrylamide gel electrophoresis gels dyed with ethidium bromide:This method, although cheap and easily accessible, has a low sensitivity and thus more ADO will be apparent than with more sensitive methods. Ethidium bromide is toxic and special care is necessary for its disposal.
2. Detection on automated sequencers using special software for fragment analysis: The most widely used is the ABI series (ABIAvant with 4 capillaries or ABI3100 with 16 capillaries). Because all these systems come with full training and support, no further discussion is necessary here.

### 3.3.8. Mutation/Polymorphism Detection

If the PCR generates fragments of different lengths, simple fragment analysis suffices. Single-nucleotide changes can be detected through several methods:

1. Restriction enzyme analysis (restriction fragment length polymorphism): Restriction sites are not always naturally present and must then be introduced with mismatch primers, further complicating the PCR.
2. Minisequencing: This method is currently widely used in PGD because it is reliable, fast, and requires only a minimum of preliminary development.
3. Real-time PCR: This method can also distinguish heterozygote and homozygote genotypes.

When working at the single cell level, nonspecific product detection and high levels of background can mask the true result from the single cell of interest. Several steps can be taken to determine the cause of the background and reduce or eliminate it:

1. For multiplex PCRs, check if background and nonspecific peaks occur when the primers are used alone. If not, this may indicate that the primers are interacting, and redesign of the primers may be necessary.
2. Adjust one of the following: annealing temperature, DMSO concentration, dNTP concentration (only when regular Taq DNA polymerase is used), or $MgCl_2$ concentration.

## 4. Notes

1. Contamination of a single cell PCR with extraneous DNA leads to serious misdiagnosis and thus must be avoided at all costs. In our experience, it usually occurs with operators in training who, after the first successful runs of single cell PCR, slacken their attention and end up with massive contamination. Occasional contamination in experienced hands is more dangerous, and it must always be kept in mind when a diagnosis is made, especially when one of the blanks is contaminated. This contamination does not indicate contamination in the concordant cell but must draw the attention to all the results in that particular PCR run. There are two types of contamination: contamination with cellular material (e.g., a cumulus cell stuck to the blastomere) and carryover contamination. The following measures deal with both types:
    a. Regarding strict separation of pre- and post-PCR laboratories, preferably, a single directional work flow from the pre- to the post-PCR facilities should be organized.
    b. The pre-PCR work should be carried out in laminar flow hoods equipped with an ultraviolet (UV) light that is turned on whenever the flow is not in use. These UV lights must be changed regularly. The flow hoods must be cleaned weekly with a soap solution, followed by ethanol.
    c. All glassware and plasticware are dedicated to the pre-PCR area, as well as all reagents. These are kept in dedicated refrigerators and freezers inside the pre-PCR area.
    d. Close-fitting gloves and surgical gowns must be worn at all times and changed frequently in the pre-PCR area. We have found that ill-fitting gloves are particularly problematic, especially because they make the opening of 0.2-mL PCR

tubes quite difficult. In our opinion, it is worthwhile to invest in the more expensive type of disposable glove and especially to draw the attention of trainees to the correct opening and closing of PCR tubes to avoid contact with any interior surface.

e. Filtered pipet tips are strongly recommended.

f. When cells are picked up, care must be taken to rinse the pipets with $Ca^{2+}$- and $Mg^{2+}$-free medium in between two cells.

g. The use of polymorphic linked or nonlinked markers in multiplex allows the detection of contamination in the single cell PCR tube itself.

2. If the amplification rate is <90%, the following troubleshooting steps can be taken:

a. Optimize the PCR, especially lowering the annealing temperature and adding DMSO to facilitate the PCR.

b. Check the quality of the cells used. Lymphoblasts should not stay in culture too long, washed cells are best used fresh, and blastomeres must contain a nucleus visible through the stereomicroscope.

c. Check the method of picking up cells to ensure that the operator is picking up cells and not debris and that the cells are not lost during the different washing steps.

If complete amplification failure is observed, do the following:

a. Check the lysis buffer expiry dates to ensure that the ALB is <1 wk old. Try the nonexpired ALB with other PCR assays.

b. Check the quality of the primers (especially if the primers worked before and the PCR yield has been seen to decrease steadily). Labeled primers should be resuspended on arrival in the laboratory, aliquoted, and ideally stored in a light-proof container at −80°C. Frequent freezing and thawing must be avoided, because this removes the label from the primer. Even when primers are kept under ideal conditions, their efficiency may deteriorate while in storage. Primers that are older than 6 mo should be checked in a trial run before being used in a clinical PGD. Moreover, most companies selling custom-made oligonucleotides will recommend that primers older than 6 mo not be used.

Finally, we wish to emphasize that especially in fluorescent PCR, primers are the most frequent cause of problems with PCR efficiency, and that these problems are always more acute in single cell PCR than in regular PCR. In other words, primers that are still working to satisfaction at the nanogram level can be totally ineffective at the single cell (picogram) level.

3. ADO is defined as the random nonamplification of one allele and can therefore be detected only in a heterozygous cell. ADO has possibly led to a number of misdiagnoses after PGD. The causes of ADO are not well known, but a number of factors certainly play a role: the presence of haploid cells in an otherwise diploid embryo; incomplete denaturation of the DNA strands in the first PCR cycles; insufficient lysis of the cells and, hence, insufficient removal of the proteins around the DNA strands; strand breaks in the native DNA, owing to endogenous DNases or

too rough treatment of the DNA. All these factors have led to the development of strategies to avoid or detect ADO:

  a. Take two cells from the embryo for analysis. This is especially recommended if simplex PCRs are used. Any discordance between the results in both cells should lead to the decision not to transfer the embryo.
  b. Increase the denaturation temperature in the first cycles of the PCR, e.g., to 96°C in the first 10 cycles. This is especially valid if regular Taq DNA polymerase is used; the ELT and EHF kits described already have a higher fixed recommended denaturation temperature.
  c. Use efficient lysis methods, such as the ALB method or the proteinase K/sodium dodecyl sulfate method (not described in this chapter). Boiling in water is obsolete and not acceptable by today's standards.
  d. Optimize the PCR to the full (e.g., concerning the length of the fragments; shorter fragments show less ADO) efficiency of the PCR (>90% amplification efficiency should be achieved), and optimize the DNA polymerase system used. Then test the PCR on heterozygous single cells. We usually test 50 cells and their corresponding blanks and aim for an amplification efficiency of >90%, and an ADO and a contamination rate of <5%.
  e. Use polymorphic linked markers in duplex or multiplex, which will allow the detection of ADO in any of the loci amplified. The accuracy achieved with duplex or multiplex PCRs compared with simplex PCRs is so much better that the biopsy of only one blastomere can be envisaged *(3)*. Multiplex PCR has become the "gold standard" in single cell PCR for PGD.

## References

1. Handyside, A., Kontogianni, E., Hardy, K., and Winston, R. (1990) Pregnancies from biopsied human preimplantation embryos sexed by Y-specific DNA amplification. *Nature* **344,** 768–770.
2. Sermon, K., Van Steirteghem, A., and Liebaers, I. (2004) Preimplantation genetic diagnosis. *Lancet* **363,** 1633–1641.
3. Thornhill, A. R., deDie-Smulders, C. E., Geraedts, J. P., et al. (2005) ESHRE PGD Consortium "Best Practice Guidelines for Clinical Preimplantation Genetic Diagnosis (PGD) and Preimplantation Genetic Screening (PGS)." *Hum. Reprod.* **20(1),** 35–48.
4. Ventura, M., Gibaud, A., Le Pendu, J., Hillaire, D., Gerard, G., Vitrac, D., and Oriol, R. (1988) Use of a simple method for the Epstein-Barr virus transformation of lymphocytes from members of large families of Reunion Island. *Hum. Hered.* **38,** 36–43.
5. De Vos, A. and Van Steirteghem, A. (2001) Aspects of biopsy procedures prior to preimplantation genetic diagnosis. *Prenat. Diagn.* **21,** 767–780.
6. Piyamongkol, W., Bermudez, M. G., Harper, J. C., and Wells, D. (2003) Detailed investigation of factors influencing amplification efficiency and allele drop-out in single cell PCR: implications for preimplantation genetic diagnosis. *Mol. Hum. Reprod.* **9(7),** 411–420.

# 5

## Real-Time Quantitative Polymerase Chain Reaction Measurement of Male Fetal DNA in Maternal Plasma

**Bernhard Zimmermann, Xiao Yan Zhong, Wolfgang Holzgreve, and Sinuhe Hahn**

### Summary

Cell-free fetal DNA can be detected in the blood plasma of pregnant women as early as 5 wk into pregnancy. At present noninvasive prenatal diagnosis has already begun to impact clinical practice. The established applications are for the determination of fetal sex and rhesus D blood group when the mother is rhesus D negative. Both methods are currently evaluated and standardized by a large laboratory network (the Special Non-Invasive Advances in Fetal and Neonatal Evaluation Network, or SAFE) that aims to implement widespread noninvasive prenatal diagnosis throughout the European Union. This chapter presents the basic methodology used in this noninvasive analysis.

**Key Words:** Real-time quantitative PCR; absolute quantification; noninvasive prenatal diagnosis; circulatory free fetal DNA; DNA analysis.

## 1. Introduction

In prenatal medicine, the discovery of fetal DNA in the circulation of pregnant women has opened a new field for noninvasive diagnosis and screening. Lo et al. *(1)* first described the detection of fetal DNA in the plasma of pregnant women by demonstrating the polymerase chain reaction (PCR) amplification of Y-chromosome-specific sequences in the plasma of women pregnant with a male fetus. Subsequently, DNA of fetal origin was found in the maternal circulation as early as 5 wk into pregnancy *(2)*. The fetal DNA comprises only a low percentage of the total DNA found in the plasma *(3)*. Owing to the high maternal DNA background, only loci clearly distinct from maternal sequences can be detected and quantified by real-time quantitative PCR *(4)*. Examples are the determination of fetal sex and Rhesus D status, and these are already applied in clinical diagnosis, and the knowledge of these fetal loci helps in preventing

From: *Methods in Molecular Medicine: Single Cell Diagnostics: Methods and Protocols*
Edited by: A. Thornhill © Humana Press Inc., Totowa, NJ

invasive procedures in the case of X-linked disorders or the unnecessary administration of anti D-immunoglobulins in the case of a Rhesus D– negative fetus *(5)*. In addition, quantification by real-time PCR has shown that fetal DNA levels are elevated in certain pregnancy-associated disorders, such as preeclampsia, preterm labor, and fetuses with trisomy 21 *(6,7)*. For single base-pair polymorphisms, short tandem repeats, or maternally inherited mutations, this simple approach is not applicable, because the high background of maternal DNA interferes with these PCR amplifications.

Here we present the methods for the determination of fetal sex and of fetal Rhesus D status if the mother is Rhesus D negative. Sequences clearly distinct from the maternal background are amplified and, therefore, the procedure is relatively straightforward. Following separation of plasma of pregnant women from the cellular fraction by centrifugation, the DNA from the plasma is extracted and analyzed by real-time PCR. The protocol allows a sensitivity of close to 100%.

## 2. Materials

### 2.1. Preparation of Sample

1. For blood collection use a tube containing EDTA as anticoagulant.

### 2.2. DNA Extraction

1. High pure PCR Template Preparation Kit (containing proteinase K, binding buffer, isopropanol, inhibitor removal buffer, and wash buffer, cat. no. 1 796 828; Roche).
2. Aerosol-resistant filter tips for pipetting, to minimize risk of contamination.
3. PCR-grade water.
4. Sterile 1.5-mL microcentrifuge/Eppendorf tubes.
5. Mechanical shaker.

### 2.3. Performing Real-Time PCR

1. For our real-time PCR analysis, we use an ABI PRISM 7000 Sequence Detection System (SDS7000).
2. For the PCR reactions, we use MicroAmp Optical 96-well reaction plates (cat. no. 4306737; Applied Biosystems, Rotkreuz, Switzerland) and ABI PRISM Optical adhesive covers (cat. no. 4311971).
3. The reactions are carried out with the TaqMan Universal PCR master mix (cat. no. 4304437; ABI) containing Amplitaq Gold DNA Polymerase, AmpErase UNG, dNTPs with dUTP and passive Reference 1 (ROX fluorescent dye) (*see* **Note 1**).
4. Primers are high-performance liquid chromatography (HPLC) purified (*see* **Note 1**); **Table 1** provides sequences specific for SRY, RhD, and GAPDH. They can be stored as 10-μ*M* aliquots at –20°C. Primers stored at 4°C are stable for several months (*see* **Note 2**).
5. We generally use TaqMan MGB probes (cat. no. 43160324; ABI). These are 5'-labeled with the fluorescent dye (VIC or FAM; note different dyes for multiplexing) and 3'-labeled with a Minor Groove Binder (MGB) and a nonfluorescent quencher

**Table 1**
**Primer and Probe Sequences for Real-Time Amplification of SRY,**
**RhD, and GAPDH**

| Primer/probe | | Sequence |
|---|---|---|
| SRY | Forward primer | TCCTCAAAAGAAACCGTGCAT |
| | Reverse primer | AGATTAATGGTTGCTAAGGACTGGAT |
| | MGB probe | TCCCCACAACCTCTT |
| | Probe (dual labeled) | CACCAGCAGTAACTCCCCACAACCTCTTT |
| RhD | Forward primer | CCTCTCACTGTTGCCTGCATT |
| | Reverse primer | AGTGCCTGCGCGAACATT |
| | MGB probe | ACGTGAGAAACGCTC |
| | Probe (dual labeled) | TACGTGAGAAACGCTCATGACAGCAAAGTCT |
| GAPDH | Forward primer | CCCCACACACATGCACTTACC |
| | Reverse primer | CCTAGTCCCAGGGCTTTGATT |
| | MGB probe | TAGGAAGGACAGGCAAC |
| | Probe (dual labeled) | AAAGAGCTAGGAAGGACAGGCAACTTGGC |

(*see* **Note 3**). It is advisable to store 5-μ*M* aliquots at –20°C, at which they are stable for more than 1 yr. If used within 1 mo, the probes can be stored at 4°C. Avoid exposure to light. **Table 1** provides probe sequences of both formulations.

6. For calibration curves, we use reference DNA of known concentration in serial dilution. The DNA is extracted from blood or buffy coat and the concentration is determined with a spectrophotometer. Dilutions are made with water or elution buffer. Store as aliquots at –20°C. Open aliquots can be used for 1 mo if stored at 4°C.

## 2.4. Analysis of Real-Time PCR Data

1. ABI Prism 7000 SDS Software.

## 3. Methods

### 3.1. Collection of Sample and Separation of Plasma

1. Collect 5 mL of peripheral maternal blood into EDTA tubes and process within 24 h. Store at room temperature or 4°C.
2. Centrifuge at 1600*g* for 10 min at room temperature or 4°C.
3. Transfer the plasma into 2-mL tubes. Be careful not to transfer any cells from the pellet.
4. Centrifuge at 16,000*g* for 10 min at room temperature.
5. Aliquot 900 μL of plasma into Eppendorf microcentrifuge tubes and store at –80°C (*see* **Note 4**).

### 3.2. DNA Extraction from Plasma

1. After thawing, spin down the plasma at 16,000*g* for 5 min and transfer 800 μL into a 2-mL tube.

2. Add 800 µL of binding buffer and 80 µL of proteinase K (50% proteinase K of the original protocol).
3. Mix and incubate for 10 min at 72°C on a shaker at low speed (400 rpm).
4. Add 400 µL of isopropanol and mix well.
5. Pipet 600 µL of the mixture into the filter supplied with the extraction kit, centrifuge at 8000*g*, and repeat until all the mixture is loaded.
6. Add 500 µL of inhibitor removal buffer and centrifuge at 8000*g*.
7. Wash two times with 500 µL of wash buffer, and centrifuge at 8000*g*.
8. Centrifuge at 13,000*g* for 10 s to remove all of the wash buffer.
9. Insert the filter into a 1.5-mL Eppendorf tube, put into the shaker at 70°C, and shake at low speed (400 rpm).
10. Add 100 mL of elution buffer or water (prewarmed to 70°C) and keep in the shaker for 1 min (*see* **Note 5**). Elute the DNA by centrifuging at 8000*g* in an Eppendorf tube (*see* **Notes 6** and **7**). One microliter of the extract will contain DNA from 8 µL of plasma.
11. Store the DNA preferably at 4°C and perform PCR within 1 wk. Freeze at –20°C if longer storage periods are required. Note that repeated freeze-thawing leads to degradation and loss of DNA.

### *3.3. Performing Real-Time PCR*

1. Include at least one negative control DNA sample from plasma in the analytic run (*see* **Note 8**).
2. Include standard DNA for the generation of a calibration curve (*see* **Note 9**). The standard DNA for the calibration curve serves as a positive control.
3. Analyze each sample in triplicate.
4. Prepare the real-time PCR reactions on ice (*see* **Note 1**).
5. Carry out the real-time PCR amplification in a total volume of 25 µL (*see* **Note 10**) containing 300 n$M$ of each primer and 200 n$M$ of each probe (*see* **Note 11**) at a 1X concentration of the TaqMan Universal PCR master mix (*see* **Note 12**).
6. First pipet the PCR reaction mixture into the wells; then add 5 µL of the DNA sample. Carefully seal the reaction plate with the optical adhesive cover. Centrifuge at 1400*g* at 4°C for 1 min to spin down any droplets and remove air bubbles.
7. Immediately start the real-time PCR with the emulation mode off (*see* **Note 13**).
8. Following an initial incubation at 50°C for 2 min, to permit AmpErase activity, and 10 min at 95°C, for activation of AmpliTaq Gold and denaturation of the DNA, use the following cycle conditions: 50 cycles of 1 min at 60°C and 15 s at 95°C.

### *3.4. Analysis of Real-Time PCR Data*

1. Perform the experimental analysis with the automatic baseline setting.
2. Set the threshold manually to a value where the signal increases exponentially for all amplifications.
3. Check all amplification curves individually by sight to identify nonexponential signals that cross the threshold.

4. For qualitative PCR, the sample sex or Rhesus D status is determined if all three replicates are positive or negative. If only one or two replicates are positive, reanalyze the sample.
5. For quantitative results, determine the copy numbers automatically from the standard curve using the SDS software. However, if one or two wells are negative, do not include them in the calculation. Thus, average independently quantities determined for replicate wells (e.g., using an Excel spreadsheet; *see* **Note 14**).
6. Determine the coefficient of variation by dividing the standard error by the average quantity of a sample. This variation will usually be in a range between 5 and 25% and can be even higher for low template numbers.

## 4. Notes

1. It is important to avoid the amplification of any nonspecific products (e.g., primer dimers and misprimed sequences) in order to avoid both false positive and false negative results. Experimental preconditions to ensure optimal specificity are HPLC-purified primers, reaction setup on ice, use of a hot start polymerase, use of AmpErase UNG, and singleplex reaction setup.
2. GAPDH serves as a control for successful DNA extraction and measures the total DNA in plasma. Unless accurate quantification of total DNA is required, one total DNA control per sample is sufficient.
3. MGB-modified probes generate a strong signal, because the quenching of the fluorescence in the intact probe is very efficient owing to its short length. This is advantageous in identifying curves from nonspecific probe cleavage with a nonexponential increase in fluorescence. Alternatively, dual labeled probes can be used (5N labeled with a fluorescent dye [VIC or FAM] and 3N labeled with a suitable quencher [TAMRA or nonfluorescent quencher]).
4. Do not refreeze plasma, because this may lead to loss of DNA in the sample.
5. DNA is more stable in the elution buffer provided by the DNA extraction kit (10 m*M* Tris buffer, pH 8.5) than in water. This effect is observed only if the DNA is stored for several months. Using water for the DNA elution allows higher input volumes into the PCR.
6. The filter was designed for elution volumes of 200 μL. Because the fetal DNA is present at very low copy numbers in the plasma of pregnant women, the elution volume is in general reduced to 50–100 μL. This allows one to obtain a higher concentration of fetal DNA in the eluate and higher detection rates close to 100%. To ensure optimal recovery of the DNA with this reduced volume, it is important to place the elution buffer onto the center of the filter. Allow the buffer 1 min to spread out on the whole filter, such that all retained DNA will go into the solution. The buffer needs to be at an elevated temperature for optimal DNA solubility. As a consequence of the low volume, the filter also needs to be prewarmed.
7. DNA recovery can be suboptimal when using reduced extraction volumes, especially 50 μL. This can be examined by eluting the filter for a second time into a *new* Eppendorf tube (with a 100-μL elution volume) and quantifying this second eluate by real-time PCR.

8. DNA extracted from purely female, Rhesus D–negative plasma is an optimal negative control. In this way, contamination of the reaction mix and also nonspecific amplification from the background DNA can be excluded.

9. Standard DNA is usually amplified in 5-fold dilutions, but 2- to 10-fold differences are also suitable. The number of reactions used for the standard curve depends on the objective of the experiment: If the purpose is only a qualitative assessment of sex or Rhesus D status, two single replicates of different standard copy numbers can serve as positive controls and simultaneously allow a crude estimation of copy number. If the aim is accurate quantification of copy numbers, duplicate or triplicate amplifications of three to six known template numbers should be performed. The use of more standard samples generally results in a lower interexperiment variation. At template numbers of fewer than 20 copies, the variability of the measurements is increased. Additionally, a reduced efficiency can often be observed owing to a lag in the first PCR cycles. Consequently, to ensure accurate quantification, standards (and samples) should have higher template numbers than 20 copies.

10. Reaction volumes between 10 and 50 µL can be used on the SDS7000. A volume of 50 µL is necessary only if a large sample input volume is needed.

11. Probe concentrations of 100 µ*M* also work well; however, the fluorescence signal will be weakened.

12. Reactions should be performed in singleplex (i.e., only one primer/probe pair per reaction). Especially GAPDH should not be multiplexed with SRY or RhD in the same tube, because the levels of total DNA are 20-fold greater than the fetal fraction. The sensitivity of the fetal-specific reactions would be markedly reduced by such multiplex analysis.

13. Run the assay always at the same heating and cooling rates. We use the emulation off setting because it is faster. The emulation mode mimics the slower heating and cooling rate of the predecessor instrument, the SDS7700.

14. To compensate for the scarcity of fetal DNA, the plasma volume used for extraction has to be increased about 10-fold compared with the protocol described here to achieve optimal results.

## References

1. Lo, Y. M., Corbetta, N., Chamberlain, P. F., Rai, V., Sargent, I. L., Redman, C. W., and Wainscoat, J. S. (1997) Presence of fetal DNA in maternal plasma and serum. *Lancet* **350,** 485–487.

2. Rijnders, R. J. P., Van Der Luijt, R. B., Peters, E. D. J., Georee, J. K., Van Der Schoot, C. E., Ploos Van Amstel, J. K., Christiaens, G. C. M. L. (2003) Earliest gestational age for fetal sexing in cell-free maternal plasma. *Prenat. Diagn.* **23,** 1042–1044.

3. Zhong, X. Y., Laivuori, H., Livingston, J. C., Ylikorkala, O., Sibai, B. M., Holzgreve, W., and Hahn, S. (2001) Elevation of both maternal and fetal extracellular circulating deoxyribonucleic acid concentrations in the plasma of pregnant women with preeclampsia. *Am. J. Obstet. Gynecol.* **184,** 414–419.

4. Hahn, S., Zhong, X. Y., Burk, M. R., Troeger, C., and Holzgreve, W. (2000) Multiplex and real-time quantitative PCR on fetal DNA in maternal plasma: a comparison with fetal cells isolated from maternal blood. *Ann. N Y Acad. Sci.* **906,** 148–152.
5. Rijnders, R. J. P., Christiaens, G. C., Bossers, B., van der Smagt, J. J., van der Schoot, E., de Haas, M. (2004) Clinical applications of cell-free fetal DNA from maternal plasma. *Obstet. Gynecol.* **103,** 157–164.
6. Zhong, X. Y., Holzgreve, W., and Hahn, S. (2001) Circulatory fetal and maternal DNA in pregnancies at risk and those affected by preeclampsia. *Ann. N Y Acad. Sci.* **945,** 138–140.
7. Lo, Y. M., Lau, T. K., Zhang, J., et al. (1999) Increased fetal DNA concentrations in the plasma of pregnant women carrying fetuses with trisomy 21. *Clin. Chem.* **45,** 1747–1751.

# 6

## Cell-Free Fetal DNA Plasma Extraction and Real-Time Polymerase Chain Reaction Quantification

### Jill L. Maron, Kirby L. Johnson, and Diana W. Bianchi

### Summary

Isolation, quantification, and genetic analysis of circulating plasma DNA have clinical applications in prenatal diagnosis, oncology, organ transplantation, posttrauma monitoring, and infectious disease. Recent technology has allowed the rapid isolation and purification of DNA from whole blood, plasma, serum, buffy coat, tissues, stool, and urine. With the advent of real-time polymerase chain reaction (PCR) amplification, extracted DNA not only can be easily identified to aid in clinical diagnoses, but also can be readily *quantified* to analyze ongoing clinical dynamics and aid in the medical prognoses of patients. Historically, identification of unique cell-free fetal DNA sequences has relied on the detection of paternally specific Y chromosome sequences owing to their relative ease in identification. However, any DNA sequence that is unique to the fetus has the potential to be amplified and quantified using real-time PCR. Our laboratory specializes in extraction of fetal DNA from maternal plasma with subsequent quantification with real-time PCR of paternally inherited sequences, such as the Y chromosome gene, *SRY*. The successful isolation and quantification of this DNA from plasma is dependent on three distinct protocols: plasma harvesting from whole blood, DNA extraction from cell-free plasma, and real-time PCR amplification and quantification of the *SRY* sequence.

**Key Words:** Cell-free fetal DNA; prenatal diagnosis; real-time quantitative polymerase chain reaction; *SRY*; *β-globin*.

## 1. Introduction

*In utero* transfer of fetal nucleated cells and, more recently, cell-free fetal DNA to a pregnant woman is well recognized *(1–7)*. Since their discovery in maternal blood, fetal nucleated cells and cell-free fetal DNA has had the potential to revolutionize the prenatal diagnoses of both fetal and maternal complications of pregnancy. While much of the early research within the field focused on separating and identifying fetal cells within maternal blood, the chances of successfully identifying the estimated 1 fetal cell/mL of maternal blood has

From: *Methods in Molecular Medicine: Single Cell Diagnostics: Methods and Protocols*
Edited by: A. Thornhill © Humana Press Inc., Totowa, NJ

limited the feasibility of fetal cells as a potential noninvasive prenatal marker *(7)*. However, in recent years cell-free fetal DNA isolated from maternal plasma or serum has offered an alternative, more practical approach to providing invaluable information about the pregnancy.

While the discovery of extracellular nucleic acids in the peripheral circulation dates back to 1948, when Mandel and Métais first observed DNA and RNA in the plasma of healthy and sick individuals *(8)*, cell-free fetal DNA was not identified in maternal plasma until 1997 *(4)*. Utilizing Y chromosomal sequences, Lo et al. *(19)* successfully identified fetal DNA in 70–80% of plasma or serum samples from women bearing male fetuses, but in none of the women carrying female fetuses. Utilizing the 5N to 3N exonuclease activity of Taq polymerase in a real-time quantitative polymerase chain reaction (PCR), Lo et al. *(9)* showed an estimated maternal plasma fetal DNA concentration of 3.4% (range: 0.39–11.9%) from 11 to 17 gestational weeks and 6.2% (range: 2.33–11.4%) late in the third trimester. Since 1997, it has been shown that fetal DNA levels in maternal plasma or serum are elevated in preeclampsia, aneuploidy, polyhydramnios, and preterm labor *(10–17)*, clearly illustrating the potential diagnostic value of fetal DNA levels in noninvasive prenatal medicine.

Although promising, the identification and quantification of cell-free fetal DNA in maternal plasma is mostly restricted to women bearing male fetuses. To identify fetal DNA properly, one must identify unique sequences that the fetus has inherited from the father, the easiest being the Y chromosome. Although some diagnostic tests have been developed to identify paternally inherited genes, including Rhesus D *(18)*, cell-free fetal DNA analysis as a diagnostic tool remains limited in its scope until the development of a genetic gender-independent DNA marker.

Successful isolation and quantification of cell-free fetal DNA from maternal plasma is dependent on three specific protocols: (1) plasma harvesting from whole blood samples; (2) DNA extraction from harvested plasma; and (3) real-time PCR quantification of a paternally inherited sequence, such as the Y chromosome sequence, *SRY*. Total DNA must also be quantified as a control to ensure proper DNA extraction from plasma. The ubiquitous *β-globin* gene is a common internal control.

Following plasma harvesting, DNA is extracted in a series of steps that ensures inhibition of nuclease activity, to prohibit DNA degradation; deproteinization of the plasma sample, to allow for accurate quantitation and hybridization of DNA; precipitation of DNA with ethanol; and, finally, elution of the available cell-free DNA for further analysis. There are several commercially available kits that provide prepared solutions and digestive enzymes that allow for easy DNA extraction.

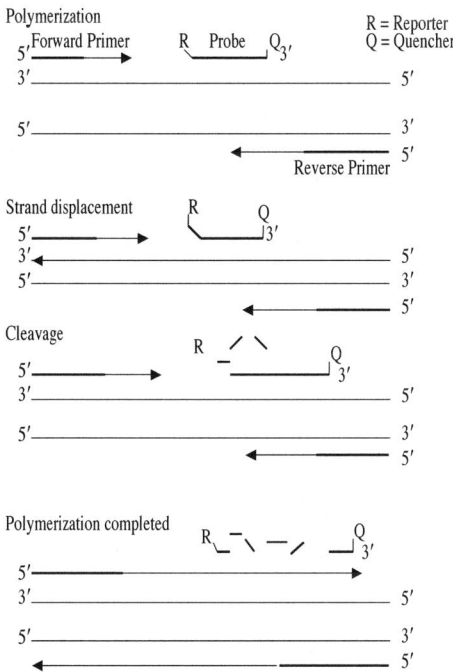

Fig. 1. Concepts of real-time PCR. (Taken from TaqMan Gold RT-PCR kit protocol [PE Applied Biosystems].)

The final step in plasma cell-free DNA quantification involves real-time PCR. Unlike standard PCR, which can only replicate a particular segment of DNA, real-time PCR utilizes oligonucleotide primers and fluorescently labeled probes not only to *amplify* DNA sequences, but also to *quantify* the DNA in a particular sample. Real-time PCR monitors PCR product formation continuously during the PCR by means of a PCR primer and a specially designed, nonextendable, fluorogenic probe. This oligonucleotide probe is dually labeled with a reporter dye covalently attached at the 5N end and a quencher dye covalently attached at the 3N end. The proximally located quencher absorbs the emission of the reporter dye as long as the probe is intact. The hybridized probe is then hydrolyzed by the 3N-nuclease activity of the TaqMan™ DNA polymerase, separating the quencher from the reporter. This results in an increase in fluorescence emission of the reporter dye, which becomes *quantitative* for the initial amount of template once exponential growth occurs above baseline threshold values. The increase in fluorescence signal is detected only if the target sequence is complementary to the probe and is amplified during PCR *(19)* (*see* **Fig. 1**). Thus, by providing quantitative analysis of extracted DNA, real-time

PCR technology can provide invaluable information on ongoing clinical dynamics in a multitude of medical situations.

## 2. Materials

### 2.1. Plasma Harvesting

1. EDTA blood collection tubes (8–10 mL).
2. Microcentrifuge tubes (1.5 mL) (P/N: 1615-5500; USA Scientific).
3. Pipet (1000 mL) (Gilson Pipette P1000, P/N: F123602; Ananchem).
4. Pipet tips with aerosol barrier (aerosol, presterilized, resistant tips; Molecular Bio-Products).

### 2.2. DNA Extraction

1. Stratalinker® 2400 Crosslinker,100v (P/N: 400676; Stratagene).
2. Qiagen QIAamp DNA Mini Blood Kit (P/N: 51304; Qiagen). The kit includes the following reagents and supplies:
   a. QIAamp Spin Columns.
   b. Collection tubes (2 mL).
   c. Buffer AL.
   d. Buffer ATL.
   e. Buffer AW1, provided as a concentrate.
   f. Buffer AW2, provided as a concentrate.
   g. Buffer AE (for elution of DNA).
   h. Qiagen® protease.
   i. Protease solvent.
   j. Proteinase K.
All reagents are stored at room temperature.
3. Microcentrifuge tubes (1.5 mL) (P/N: 1605-5500; USA Scientific).
4. Microcentrifuge tubes (0.5 mL) (P/N: 1605-4400; USA Scientific).
5. Pipet tips with aerosol barrier (aerosol, presterilized, resistant tips; Molecular Bio-Products).
6. 100% Ethanol (for precipitation of DNA).

### 2.3. Real-Time PCR

1. *SRY* primers (store at –80°C):
   a. Forward: 5NTCCTCAAAAGAAACCGTGCAT 3N.
   b. Reverse: 5N AGATTAATGGTTGCTAAGGACTGGAT 3N.
2. SRY probe (store at –20°C): 5N (6-FAM)-CACCAGCAGTAACTCCCCACAAC-CTCTTT-(TAMRA) 3N. The probe is light sensitive and must be stored and thawed in a light-blocking container.
3. *β-globin* primers (store at –80°C):
   a. Forward: 5NGTGCACCTGACTCCTGAGGAGA 3N.
   b. Reverse: 5NCCTTGATACCAACCTGCCCAG 3N.

4. β *globin* probe (store at –20°C): 5N (6 FAM) AAGGTGAACGTGGAT GAAGTTGGTGG-(TAMRA) 3N. The probe is light sensitive and must be stored and thawed in a light-blocking container.
5. TaqMan® Universal PCR Master Mix (store at 4°C) (P/N: 4304437) PE Applied Biosystems.
6. Amber microcentrifuge tubes (1.5 mL) (P/N:1615-5507; USA Scientific).
7. Amber microcentrifuge tubes (0.5 mL) (P/N: 1615-0007; USA Scientific).
8. 98-Well Optical Reaction Plate (P/N: 4306737; Applied Biosystems).
9. Optical caps (P/N: 4323032; Applied Biosystems).
10. Diethylpyrocarbonate (DEPC)-treated water.

## 3. Methods

The methods described outline the specific protocols necessary for cell-free fetal DNA extraction and subsequent quantification. For the purpose of this chapter, the amplification and quantitation of the Y chromosome–specific sequence, *SRY*, is described. To demonstrate that DNA was successfully isolated from each maternal sample regardless of fetal gender, the amplification and quantitation of the ubiquitous housekeeping β*-globin* DNA sequence, as a marker of total DNA, is also described.

### 3.1. Plasma Harvesting

The total volume of starting plasma necessary for successful cell-free fetal DNA extraction varies with gestational age. The level of fetal DNA in the maternal circulation increases steadily during the first and second trimesters, with a sharp increase during the last 8 wk of the third trimester (*9*). We therefore recommend a minimum of 800 µL of starting plasma for first-trimester samples and a minimum of 400 µL of starting plasma for second- and third-trimester maternal samples (*see* **Note 1**).

1. Using a sterile phlebotomy technique, obtain 8–10 cc of whole blood from the subject and place immediately in EDTA tubes.
2. Allow the blood in the tubes to equilibrate to room temperature (approx 1 h) while gently inverting on a rocker.
3. Centrifuge the blood in its original vacuum container at 1000 rpm (200*g*) for 10 min at room temperature.
4. Under a sterile hood, divide the plasma into 1.0-mL aliquots and place into 1.5-mL microcentrifuge tubes using aerosol-free pipet tips. Take care not to disturb the cellular fraction. Leave approx 1 mL of plasma above the cells.
5. Centrifuge aliquoted plasma samples at 14,000 rpm (15,339*g*) in a microcentrifuge for 10 min to remove any residual cells. Our laboratory recommends performing the centrifugation at 4°C to prevent the rotor from overheating, which can lead to degradation of DNA and limit its extraction.
6. Using sterile, filtered, aerosol-free pipets, carefully transfer 900 µL of newly centrifuged plasma into fresh 1.5-mL labeled Eppendorf tubes. Now 900-µL

samples may be stored at –80°C for further analysis or directly processed for DNA extraction.

### 3.2. Extraction of Cell-Free Fetal DNA Plasma

In our laboratory, we use the Qiagen QIAamp DNA Mini Kit™ for extraction of DNA plasma. We have modified its "Blood and Body Fluid Spin Protocol" slightly to achieve maximum DNA yield. The protocol outlined next uses as a starting volume 400 μL of plasma. As stated previously, larger volumes of 800 μL may be needed, particularly for first-trimester DNA samples. In such instances, the protocol may be modified further to handle the larger volume. Those modifications are provided under **Subheading 3.2.3.**

### 3.2.1. Preparation

Prior to starting extraction, perform the following steps:

1. Place all pipets and microcentrifuge tubes needed for extraction (one 0.5-mL and two 1.5-mL microcentrifuge tubes per sample) into an ultraviolet (UV) crosslinker for 20 min. This step only serves to UV crosslink DNA, prohibiting its amplification and reducing exogenous contamination.
2. Heat a water bath to 56°C.
3. Equilibrate the samples to room temperature.
4. Prepare Buffer AW1 and Buffer AW2, supplied in the QIAamp kit, per the manufacturer's protocol.
5. If frozen, thaw Qiagen protease at room temperature.
6. Defrost the plasma sample at room temperature.

### 3.2.2. DNA Extraction

1. Add 400 μL of prepared plasma sample into a 1.5-mL microcentrifuge tube.
2. Add 40 μL of Qiagen protease and mix well by pulse vortexing for 15 s.
3. Add 400 μL of Buffer AL to the sample. Mix thoroughly by pulse vortexing for 15 s. Do *not* add Qiagen protease directly to Buffer AL.
4. Incubate in the 56°C water bath for 10 min. Avoid extended incubations in order to limit DNA degradation.
5. Add 400 μL of 100% ethanol to the sample and mix by pulse vortexing for 15 s.
6. Carefully apply 650 μL of the mixture prepared from **step 5** to the QIAamp spin column in a 2-mL collection tube. Be careful *not* to wet the rim of the tube. Close the cap and centrifuge the sample at 14,000 rpm (15,339*g*) for 1 min. Place the spin column in a clean 2-mL collection tube. Repeat until the entire sample is loaded onto the column (approx two times).
7. Carefully open the spin column and add 500 μL of Buffer AW1 without wetting the rim. Close the cap and centrifuge at 14,000 rpm (15,339*g*) for 1 min. Place the spin column in a clean 2-mL collection tube.
8. Carefully open the spin column and add 500 μL of Buffer AW2 without wetting the rim. Close the cap and centrifuge at 14,000 rpm (15,339*g*) for 3 min.

9. Place the spin column in a new 2-mL collection tube and spin dry at 14,000 rpm (15,339$g$) for 3 min. This step removes any residual Buffer AW2 that may interfere with the subsequent PCR procedure.

10. Place the spin column into a labeled 1.5-mL microcentrifuge tube. Carefully open the spin column and add 50 µL of Buffer AE to the center of the column. It is essential that the pipet tip not directly touch the filter. Incubate the spin column loaded with Buffer AE for 5 min at room temperature. Centrifuge at 14,000 rpm (15,339$g$) for 1 min.

11. Open the spin column and add another 50 µL of Buffer AE to the center of the column. Incubate the spin column for 5 min and centrifuge at 14,000 rpm (15,339$g$) for 1 min.

12. Discard the column and store the elution at 4°C until immediate PCR amplification or at –20°C until future analysis.

### 3.2.3. Modifications for 800 µL Starting Plasma Volume

1. UV irradiate one 5-mL Falcon polypropylene tube in a Stratalinker in lieu of a smaller 1.5-mL microcentrifuge tube to handle greater volume.

2. Add 80 µL of Qiagen protease to the original 800-µL plasma sample in the 5-mL Falcon polypropylene tube.

3. Add 800 µL of Buffer AL to the sample and mix thoroughly by pulse vortexing for 15 s.

4. After incubating at 56°C for 10 min, add 800 µL of 100% ethanol to the sample and mix again by pulse vortexing for 15 s. No other modifications are needed after this step. Proceed with **steps 6–12** in **Subheading 3.2.2**.

## 3.3. TaqMan Real-Time PCR DNA Quantification of SRY and β-globin Sequences

TaqMan provides a specifically designed Universal PCR Master Mix that contains AmpliTaq Gold DNA Polymerase, a thermal stable DNA polymerase that has 5N-3N nuclease activity but lacks 3N-5N exonuclease activity; AmpErase UNG, to prevent contamination of sample with carryover DNA; dNTPs with dUTP; Passive Reference, an internal control to which the reporter-dye signal can be normalized during data analysis; and buffers. The Universal PCR Master Mix is meant to be used with any appropriately designed DNA-specific forward and reverse primers and a fluorescently labeled probe. When amplifying and quantifying any target DNA sequence, it is imperative to also amplify and quantify a known DNA sequence control to ensure that DNA was indeed present in the extracted sample (*see* **Note 2**). In our laboratory, the ubiquitous DNA *β-globin* sequence is analyzed in a separate real-time PCR reaction to ensure that DNA was successfully extracted from each plasma sample. All samples should be run in triplicate, and the standard curve should be run in duplicate. A no-template control consisting of 5 µL of DEPC water should be run with each plate (*see* **Note 3**).

### 3.3.1. Preparation of Working Stocks of Primers and Probes

*SRY* and *β-globin* primers and probes may be obtained from various vendors. We recommend resuspending the primers and probes to achieve the following optimized concentrations for PCR:

| Sequence | Forward primer | Reverse primer | Probe |
|----------|----------------|----------------|-------|
| SRY | 100 nm | 100 nm | 100 nm |
| *β-globin* | 200 nm | 200 nm | 50 nm |

New lots of primers and probe should be assayed in parallel with current working stock solutions using standard curve DNA as template. Quality control criteria for new solutions should include linearity of the standard curve, standard curve slope, and count values to monitor fluorescence intensity.

### 3.3.2. Preparation of Standard Curves

Prepared standard curve solutions for various DNA sequences may be obtained from multiple vendors. Alternatively, one may isolate a known quantity of the DNA sequence of interest and serially dilute the sample to obtain a linear standard curve. It is imperative, however, that the dilutions for the standard curve have an adequate range to be able to successfully quantify the amount of cell-free DNA in the unknown samples. For *SRY* and *β-globin* standard curves, our laboratory has found that a range of values from 6.4 pg/5 µL to 20,000 pg/5 µL is sufficient to properly identify the unknown quantitative values of the extracted cell-free fetal DNA. We recommend, however, that individual laboratories empirically determine an adequate range of values for each probe being analyzed. Strict correlation coefficients (e.g., >0.97) and line slopes (e.g., −3.3) should also be adhered to.

### 3.3.3. PCR Plate Setup (SRY and β-globin)

Each well in the plate will contain a total volume of 50 µL consisting of the following reagents: 20 µL of Universal PCR Master Mix, 20 µL of dH$_2$O, 1.25 µL of forward primer (*SRY* or *β-globin*), 1.25 µL of reverse primer (*SRY* or *β-globin*), 2.5 µL of probe (*SRY* or *β -globin*), and 5 µL of sample DNA.

1. Prior to the setup of the plate, calculate the number of samples to be run in triplicate.
2. Multiply this calculated number of wells by each of the volumes listed in the preceding paragraph to determine how much reagent will be needed for each plate.
3. UV irradiate three 0.5-mL and three 1.5-mL amber microcentrifuge tubes to be used in reagent preparation for 20 min in a Stratalinker. Amber tubes are utilized to block any excess exogenous UV light that may degrade the probe.

| | 1 | 2 | 3 | 4 | 5 | 6 | 7 | 8 | 9 | 10 | 11 | 12 |
|---|---|---|---|---|---|---|---|---|---|---|---|---|
| **A** | 20000 | 4000 | 800 | 160 | 32 | 6.4 | dH$_2$O | dH$_2$O | dH$_2$O | | | |
| **B** | 20000 | 4000 | 800 | 160 | 32 | 6.4 | | | | | | |
| **C** | | | | | | | | | | | | |
| **D** | | | | | | | | | | | | |
| **E** | | | | | | | | | | | | |
| **F** | | | | | | | | | | | | |
| **G** | | | | | | | | | | | | |
| **H** | | | | | | | | | | | | |
| | 1 | 2 | 3 | 4 | 5 | 6 | 7 | 8 | 9 | 10 | 11 | 12 |

Fig. 2. Example of a 96-well PCR plate. Wells A and B 1–6 represent a known standard curve (pg/5 μL of DNA stock solution) of DNA probe, run in duplicate. Wells A 7–9 serve as the no-template control run in triplicate. The remaining wells should be used to run unknown samples in triplicate.

4. Retrieve reagents from respective storage conditions and place on ice. Remember to keep the probe in a light-blocking container.
5. Combine Universal PCR Master Mix and dH$_2$O in the 1.5-mL amber microcentrifuge tubes and gently mix.
6. Combine the forward primer, reverse primer, and probe in 0.5-mL amber microcentrifuge tubes and gently mix.
7. Add the primer and probe solution directly to the Universal PCR Master Mix and dH$_2$O and gently mix.
8. Retrieve the 96-well plate to prepare for setup.
9. In each well, including the standard curve and control wells, pipet 45 μL of prepared master mix, dH$_2$O, and primer and probe solutions (*see* **Fig. 2** for an example of a plate setup).
10. Add 5 μL of standard curve, DNA sample, or dH$_2$O to each respective well.
11. Securely cover the plate with optical caps.
12. Centrifuge at 1400 rpm (400g) for 5 min at room temperature (with the break off). The plate is now ready for amplification and quantification with real-time PCR.

### 3.3.4. Thermal Cycling Parameters

For the purpose of *SRY* and *β-globin* DNA, we recommend the following thermal cycling parameters:

1. Denaturing: 95°C for 10 min for 1 cycle.
2. Annealing: 95°C for 15 s followed by 60°C for 1 min for a total of 40 cycles.

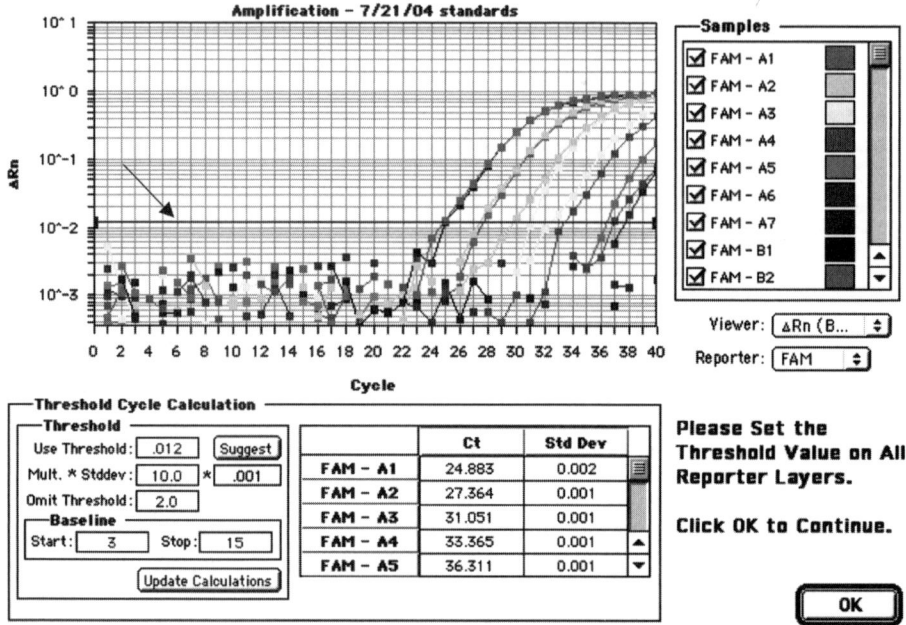

Fig. 3. Example of threshold determination. The arrow indicates where the threshold (dark horizontal line) is drawn above background fluorescence. As noted in the bottom left corner, the computer will "suggest" a threshold level (ΔRn) at which amplification of DNA is occurring above background fluorescence. Alternatively, the user may manually manipulate the dark horizontal line to perfect the image and eliminate the random scatter points representative of background fluorescence. Threshold determination can vary between reactions depending on the degree of background fluorescence.

### 3.3.5. Data Analysis

The sophisticated hardware and software of real-time PCR systems quickly and easily provide data analysis for each real-time PCR experiment. We recommend that the instructions provided with the system being used be followed carefully and that the support staff be used to troubleshoot major difficulties.

During analysis, the computer will automatically assign a level at which fluorescence begins to exceed threshold during the exponential identification of generated product. Although this is a good starting point, we recommend that one's own laboratory devise an objective means by which to determine empirically the threshold cycle. Consistency among users is essential when assigning a threshold to limit inaccuracies in data analysis (*see* **Fig. 3**).

## 4. Notes

1. Occasionally, DNA yield following extraction will not be sufficient. As stated previously, overall starting plasma volume may be a contributing factor to this prob-

lem. However, there are additional steps that can also be undertaken in an attempt
to maximize DNA yield:

    a. According to the extraction protocol, final elution of DNA from the column is
accomplished with AE Buffer. Our laboratory finds that elution with DEPC
water rather than AE Buffer may increase DNA yield. However, elution with
water is only recommended if the sample will be quantified with real-time PCR
in a timely fashion. Prolonged storage of extracted DNA at –20°C eluted with
DEPC water will hydrolyze the DNA and ruin the sample.

    b. Warming the AE Buffer to 56°C prior to the elution step may also increase DNA
yield.

    c. Concentrating the elution volume to a minimum of 35 µL may increase DNA
yield.

    d. Re-eluting the column with original eluate, incubating for 5 min at room tem-
perature, and spinning for an additional minute at 14,000 rpm may also improve
yield.

2. Failure of DNA amplification during real-time PCR will occasionally occur. Clearly,
no amplification will occur if Y chromosomal probes are attempting to be amplified
from a mother carrying a female fetus. However, excluding this as a potential culprit,
other possibilities for lack of DNA amplification include the following:

    a. Exhaustion of reaction components, specifically nucleotides: If appropriate con-
centrations of reagents are not present in each reaction well, there will be failure of DNA amplification during the exponential phase of the reaction.

    b. Inappropriate probe design: Ideally, probes should have a G-C content in the
30–80% range and not have Gs on the 5N end. In addition, select the strand that
gives the probe more Cs than Gs and avoid runs of an identical nucleotide, particularly runs with four or more Gs.

    c. Inappropriate primer design: Ideally, primers should be as close as possible to
the probe without overlapping it. Finally, the five nucleotides at the 3N end
should have not more than two G and/or C bases.

3. Contamination, either through the addition of exogenous Y chromosomal DNA or
DNase to the sample or through residual amplicon contamination during real-time
PCR serves as one of the greatest hindrances to successful isolation and quantification of DNA. However, there are several steps that can be performed to limit contamination during sample preparation:

    a. Always wear gloves and other protective clothing when handling samples, tubes,
and instruments. Although DNase is not as ubiquitous as RNase, it is still prevalent and can quickly contaminate a sample and degrade DNA. In addition, male
researchers specifically attempting to isolate Y chromosome DNA probes
should realize that they have the ability to contaminate any sample with their
own DNA, yielding false positive and/or quantitatively higher DNA results

    b. UV irradiate all tubes and pipets for 20 min prior to use in a Stratalinker. This
should eliminate residual DNA contamination. DNase Away™ (P/N: E/K-3356;
E and K Scientific) is commercially available for cleaning the equipment and
the hood prior to use. It is our experience, however, that 70–100% ethanol is
equally effective.

4. PCR contamination with amplicons from previous runs can also occur, because DNA fragments can become aerosolized and spread throughout the laboratory environment. To minimize such contamination, it is recommended that a designated PCR room be established complete with separate pipettors, pipet tips, and reaction tubes. Additionally, certain TaqMan polymerase kits contain uracil-$N$-glycosylase (UNG). This enzyme reduces contamination because it substitutes dUTP for dTTP in the PCR. By doing so, the dU-containing amplicons are destroyed by treatment with UNG. The UNG enzyme is then denatured during the initial denaturation step in the PCR so that newly synthesized PCR products are not degraded. UNG does not affect the template DNA.

## References

1. Lo, Y. M. D. (2001) Circulating nucleic acids in plasma and serum: an overview. *Ann. N. Y. Acad. Sci.* **945,** 1–7.
2. Lo, Y. M. D. (2000) Fetal DNA in maternal plasma: biology and diagnostic applications. *Clin. Chem.* **46,**1903–1906.
3. Lo, Y. M. D., Tien, M. S. C., Lau, T. K., et al. (1998) Quantitative analysis of fetal DNA in maternal plasma and serum: implications for noninvasive prenatal diagnosis. *Am. J. Hum. Genet.* **62,** 768–775.
4. Lo, Y. M. D., Corbetta, N., Chaberlain, P. F., Rai, V., Sargen, I. L., Redman, C. W., and Wainscoat, J. S. (1997) Presence of fetal DNA in maternal plasma and serum. *Lancet* **350,** 485–487.
5. Bianchi, D. W., Williams, J. M., Sullivan, L. M., Hanson, F. W., Klinger, D. W., and Shuber, A. P. (1997) PCR quantitation of fetal cells in maternal blood in normal and aneuploid pregnancies. *Am. J. Hum. Genet.* **61,** 822–829.
6. Lo, Y. M. D., Corbetta, N., Chaberlain, P. F., Rai, V., Sargen, I. L., Redman, C. W., and Wainscoat, J. S. (1997) Presence of fetal DNA in maternal plasma and serum. *Lancet* **350,** 485–487.
7. Bianchi, D. W. (1999) Fetal cells in the maternal circulation: feasibility for prenatal diagnosis. *Br. J. Haematol.* **105,** 547–583.
8. Mandel, P. and Métais, P. (1948) Les acides nucléiques du plasma sanguine chez l'homme. *Acad. Sci. Paris* **142,** 241–243.
9. Lo, Y. M. D., Tien, M. S. C., Lau, T. K., et al. (1998) Quantitative analysis of fetal DNA in maternal plasma and serum: implications for noninvasive prenatal diagnosis. *Am. J. Hum. Genet.* **62,** 768–775.
10. Lo, Y. M. D., Leung, T. N., Tien, M. S. C., Sargent, I. L., Zhang, J., Lau, T. K., and Haines, C. J. (1999) Quantitative abnormalities of fetal DNA in maternal serum in pre-eclampsia. *Clin. Chem.* **45,** 184–188.
11. Levine, R. J., Qian, C., Leshane, E. S., et al. (2004) Two-stage elevation of cell-free fetal DNA in maternal sera before onset of preeclampsia. *Am. J. Obstet. Gynecol.* **190,** 707–713.
12. Leung, T. N., Zhang, J., Lau, T. K., Hjelm, N. M., and Lo, Y. M. D. (1998) Maternal plasma fetal DNA as a marker for preterm labour. *Lancet* **352,** 1904, 1905.

13. Lee, T., LeShane, E. S., Messerlian, G. M., Canick J. A., Farinal, A., Heber, W. W., and Bianchi, D. W. (2002) Down syndrome and cell-free fetal DNA in archived maternal serum. *Am. J. Obstet. Gynecol.* **187,** 1217–1221.
14. Wataganara, T., LeShane, E. S., Farina, A., Messerlian, G. M., Lee, T., Canick, J., and Bianchi, D. W. (2003) Maternal serum cell-free fetal DNA levels are increased in cases of trisomy 13 but not trisomy 18. *Hum. Genet.* **112,** 204–208.
15. Farina, A., LeShane, E. S., Lambert-Messerlian, G. M., Canick, J. A., Lee, T., Neveux, L. M., Palomaki, G. E., and Bianchi, D. W. (2003) Evaluation of cell-free fetal DNA as a second-trimester maternal serum marker of Down syndrome pregnancy. *Clin. Chem.* **49,** 239–242.
16. Zhong, X. Y., Bürk, M. R., and Troeger, C. (2000) Fetal DNA in aneuploid fetuses is elevated in pregnancies with aneuploid fetuses. *Prenat. Diagn.* **20,** 795–798.
17. Zhong, X. Y., Holzgreve, W., and Li, J. C. (2000) High levels of fetal erythroblasts and fetal extracellular DNA in the peripheral blood of a pregnant woman with idiopathic polyhydramnios: case report. *Prenat. Diagn.* **20,** 838–841.
18. Turner, M. J., Martin, C. M., and O'Leary, J. J. (2003) Detection of fetal Rhesus D gene in whole blood of women booking for routine antenatal care. *Eur. J. Obstet. Gynecol. Reprod. Biol.* **108,** 29–32.
19. O'Connell, J., ed. (2002) *RT-PCR Protocols,* Humana, Totowa, NJ.

# 7

## Linear-After-The-Exponential Polymerase Chain Reaction and Allied Technologies

*Real-Time Detection Strategies for Rapid, Reliable Diagnosis from Single Cells*

### Kenneth E. Pierce and Lawrence J. Wangh

### Summary

Accurate detection of gene sequences in single cells is the ultimate challenge to polymerase chain reaction (PCR) sensitivity. Unfortunately, commonly used conventional and real-time PCR techniques are often too unreliable at that level to provide the accuracy needed for clinical diagnosis. Here we provide details of linear-after-the-exponential-PCR (LATE-PCR), a method similar to asymmetric PCR in the use of primers at different concentrations, but with novel design criteria to ensure high efficiency and specificity. Compared with conventional PCR, LATE-PCR increases the signal strength and allele discrimination capability of oligonucleotide probes such as molecular beacons and reduces variability among replicate samples. The analysis of real-time kinetics of LATE-PCR signals provides a means for improving the accuracy of single cell genetic diagnosis.

**Key Words:** Asymmetric PCR; cell lysis; fluorescent probes; molecular beacons; proteinase K; real-time PCR.

## 1. Introduction

The polymerase chain reaction (PCR) provides a method for identifying alleles of specific genes, or the mRNA transcribed from those genes. Through the 1980s and most of the 1990s, the products of PCR amplification were characterized using postamplification methods such as restriction enzyme treatment followed by electrophoresis through agarose or polyacrylamide gels. These and other postamplification detection strategies are time-consuming and increase the risk of contaminating subsequent assays. This is particularly problematic in the case of single cell samples, because a single product molecule inadvertently introduced into a sample tube can generate a false positive result and lead to a misdiagnosis.

From: *Methods in Molecular Medicine: Single Cell Diagnostics: Methods and Protocols*
Edited by: A. Thornhill © Humana Press Inc., Totowa, NJ

Real-time PCR using fluorescent probes *(1–3)* allows the kinetics of the amplification process to be observed and analyzed. Moreover, the fact that real-time PCR is carried out in closed tubes greatly reduces the risk of laboratory contamination, saves time, and is amenable to automation. Real-time assays using TaqMan™ probes have become popular for many applications, primarily owing to the "assays on demand" program from Applied BioSystems for primer and probe design. However, the TaqMan assay requires digestion of the probe by the exonuclease activity of Taq polymerase, a process that requires probes with a relatively high melting temperature ($T_m$). This, in turn, makes it more difficult to distinguish allelic variants and can also reduce amplification efficiency. Molecular beacons and several other types of commercially available probes have greater allele-discriminating capacities than TaqMan probes but have design constraints of their own.

Regardless of which type of probe is used to monitor a symmetric real-time amplification, hybridization of the probe to its target must compete with the reannealing of the complementary amplicon strands. By the end of the reaction, amplicon strand reannealing predominates and the probe detects only a fraction of the total number of amplicons produced (**Fig. 1A**). To circumvent this problem, we investigated the possible use of asymmetric PCR. Asymmetric PCR uses unequal concentrations of primers first to amplify both DNA strands exponentially, then shifts to linear amplification of one strand on depletion of the limiting primer. The DNA strand that is produced by the extension of the excess primer during the linear phase is freely accessible for hybridization to the probe (**Fig. 1B**). However, traditional asymmetric PCR that makes use of primers designed for symmetric amplifications *(4)* is inefficient, highly variable, and tends to generate high levels of nonspecific product. Those undesirable characteristics can be overcome if primers are designed for use at unequal concentrations. The resulting amplification strategy, termed Linear-After-The-Exponential PCR (LATE-PCR) is efficient and specific *(5,6)*. **Figure 2** shows a comparison of symmetric PCR and LATE-PCR for the detection of the ΔF508 allele of the cystic fibrosis gene (*CFTR*) in single cells.

LATE-PCR also makes it possible to use lower temperature detection, because the probe does not need to compete with hybridization and extension of the limiting primer during the early, exponential phase of the reaction. Hybridization of probe and target is unimpeded once the limiting primer is depleted and can be done either by lowering the annealing temperature at that point, or by introducing a low-temperature detection step between the extension and melting steps. Probes with lower melting temperatures are easier to design, more allele discriminating, and have lower background fluorescence. Moreover, because the probe dissociates from its target strand well below the extension temperature of the reaction, sufficient probe can be added to the reaction to

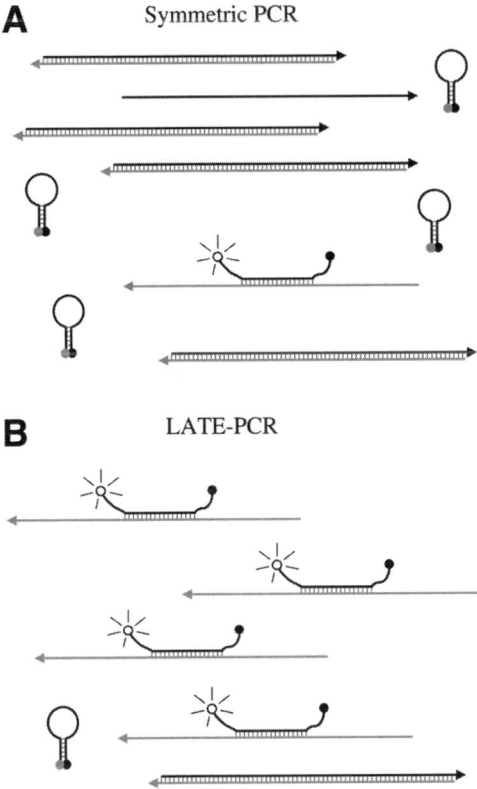

Fig. 1. Schematic comparison of symmetric PCR and LATE-PCR for amplicon detection using molecular beacons. Near the completion of symmetric PCR (**A**), the complementary strands of the amplicon (black and gray lines with arrows representing the 3' ends) reach high concentrations and reanneal. Molecular beacon molecules unable to hybridize with those targets remain in the nonfluorescent, hairpin configuration. LATE-PCR (**B**) generates an excess of the amplicon strand that is the target of the molecular beacon. Molecular beacons readily hybridize to those strands and emit fluorescence, generating a much greater total fluorescent signal from the LATE-PCR sample.

measure all product strands without inhibiting the amplification reaction (*5*). We have used these features of LATE-PCR for constructing single cell assays for several alleles of cystic fibrosis (*[7]*; unpublished data), Tay-Sachs disease (*5*), and β-thalassemia and p53 (unpublished data). Here we provide practical information for the design and use of LATE-PCR assays.

## 2. Materials

1. Cells with desired genotypes for positive controls (Coriell Cell Repositories).
2. Microscope of choice for cell analysis and transfer.

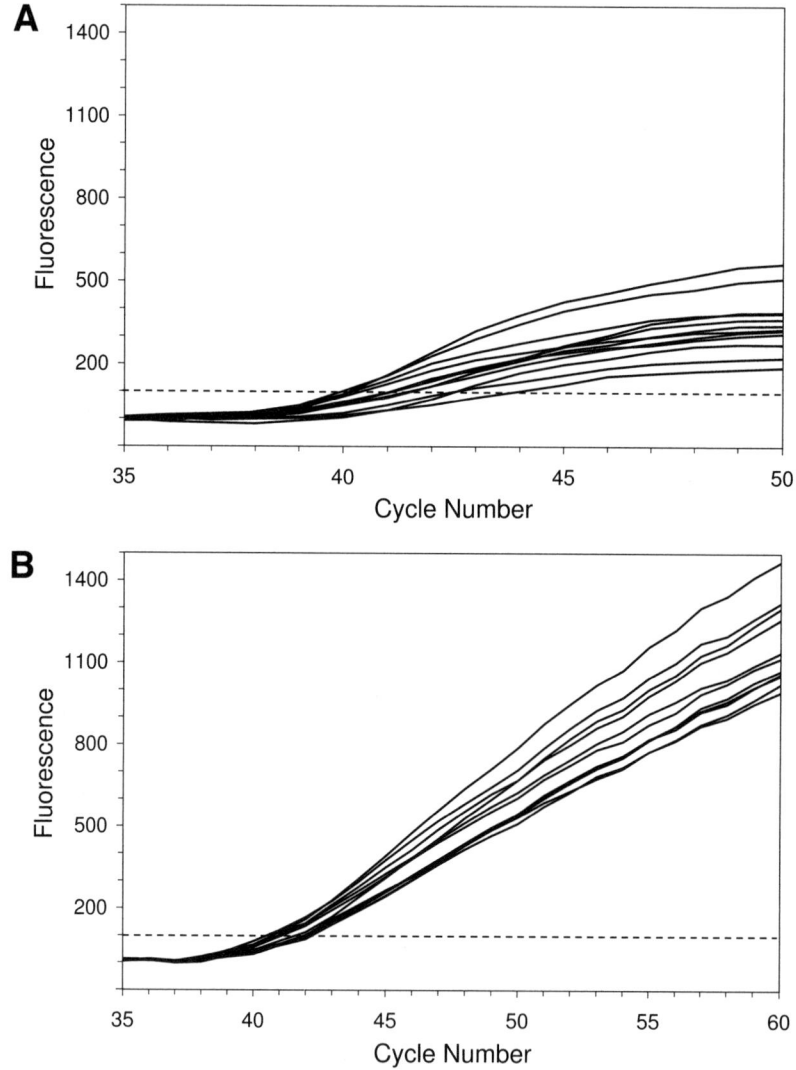

Fig. 2. Real-time PCR results for detection of ΔF508 allele in single, heterozygous lymphoblasts using molecular beacons. (**A**) Symmetric PCR replicate samples exhibit wide ranges of $C_T$ values (the point at which fluorescence reaches the dashed threshold line) and low final fluorescence. (**B**) LATE-PCR replicate samples have relatively low variation in $C_T$ values and much higher final fluorescence.

  3. PCR enclosure hoods (e.g., Labconco Purifier™).
  4. Low-attachment culture dishes (e.g., Corning 6-well; cat. no. 3471).
  5. Narrow-bore cell transfer pipets and micromanipulators.
  6. Mechanical pipettors and aerosol-resistant pipet tips.

7. Calcium-free, magnesium-free phosphate-buffered saline (PBS) (Sigma).
8. Nonacetylated bovine serum albumen (BSA) or polyvinylpyrrolidone (PVP) (optional) (Sigma).
9. Lysis solution: 100 μg/mL of proteinase K (Roche); 5 μ*M* sodium dodecyl sulfate (SDS) (Sigma); 10 m*M* Tris-HCl, pH 8.3 (TRIZMA® Pre-Set Crystals, Sigma).
10. PCR primer design software.
11. Thermal cycler with fluorescence detection capability (e.g., ABI PRISM® 7000 or 7700, Bio-Rad iCycler, Stratagene Mx3000P™ or Mx4000®, or Cepheid SmartCycler®).
12. Optical sample tubes appropriate to the thermal cycler.
13. Racks for placing sample tubes on ice (e.g., ABI MicroAmp® Bases).
14. Standard thermal cycler or heating blocks (with heated cover) for lysis reaction (optional).
15. PCR reagents:
    a. Taq polymerase with hot-start capacity (either with anti-Taq antibodies such as Platinum Taq [Invitrogen], or with modified enzyme such as AmpliTaq Gold [ABI]).
    b. Buffers containing Tris and KCl (usually supplied with commercial Taq polymerases).
    c. MgCl$_2$ stock solution at 25 or 50 m*M*.
    d. Custom oligonucleotide primers and probes.
    e. Deoxynucleotide triphosphates (dNTPs), PCR grade (Promega).
    f. Water, molecular biology grade.
    g. SYBR Green I (Molecular Probes) (optional).

## 3. Methods

Obtaining reproducible results from samples of single cells requires (1) sample preparation that avoids inhibitors of PCR and removes chromosomal proteins from the DNA; (2) the use of primers and probes that maximize amplification efficiency, specificity, and signal strength; and (3) analysis of real-time signal kinetics from tested samples and from controls with known genotypes.

### 3.1. Preparation and Lysing of the Cell

The choice of methods for isolating single cells will vary considerably depending on the cell type and available equipment. For instance, cells in suspension can be isolated individually by hand-controlled micromanipulation or by fluorescence-activated cell sorting. Alternatively, fixed or embedded cells can be isolated using laser capture microscopy, although the required equipment is expensive and not widely available. This chapter provides only general information on this topic with the intention of pointing out potential pitfalls that can affect cell lysis, genomic DNA preparation, and subsequent PCR.

## 3.1.1. Isolation and Washing of the Cell

When cell isolation is carried out manually, cells should first be diluted to a density that facilitates picking up individual cells using either a handheld pipet or a pipet in a micromanipulator. Adherent cells should be dissociated by repeated pipetting, preferably in a calcium-free, magnesium-free medium or PBS. Petri dishes or microtiter plates with low-adhesion surfaces can reduce the chances of cell damage or loss. Solution additives such as BSA or PVP can also be used for this purpose, but any additive should be carefully evaluated for its effect on cell lysis and amplification (*see* **Note 1**).

Several components present in culture medium or used in cell isolation techniques can inhibit PCR and must be removed by transferring the cell through PBS or culture medium that lacks the inhibitors (*see* **Note 2**). One or two rinses may be sufficient if the transferred volume can be kept to a minimum (e.g., overall volume dilution of 1:100 or greater per step). Transfers should be practiced before attempting to manipulate valuable, limited-source cells. First aspirate a small amount of the wash solution into the transfer pipet, and then aspirate the cell into the tip of the pipet. Carefully expel the contents of the pipet into the wash while examining under the microscope. As soon as the cell exits the pipet, remove the pipet from the wash dish, expel the remaining solution into a separate container, and rinse the pipet in unused wash solution. Repeat this procedure as necessary to reduce the concentration of potential PCR inhibitors. All washes should be brief.

## 3.1.2. Preparation of Lysis Solution and Transferring of the Cell

Using real-time detection of multicopy genes, we demonstrated that a properly buffered solution containing proteinase K and SDS provides the greatest number of targets for amplification (*8*). This lysis reagent can be prepared in advance and stored for up to at least 1 yr at −20°C in a constant-temperature freezer (i.e., not frost free). Other tested lysis methods resulted in more variable recovery and/or delayed detection, presumably owing to either DNA damage or incomplete removal of chromosomal proteins from the DNA. A delay in PCR signals can also indicate inefficient amplification owing to the presence of PCR inhibitors.

Shortly before preparing the cells, an aliquot of the lysis reagent is thawed on ice and 10 µL is pipetted to each PCR sample tube (*see* **Note 3**). It is extremely important to use procedures that minimize the likelihood of contamination when preparing or working with lysis solution and PCR reagents (*see* **Note 4**). Sample tubes should be kept on ice until the cells have been transferred, because proteinase K is self-digesting at rates that are temperature dependent. Any unused solution should be discarded, because repeated freezing and thawing may reduce enzyme activity.

The final cell transfer is done directly into lysis solution in a PCR tube (*see* **Note 5**), keeping the volume of the transferred wash solution to a minimum. Depending on the type of sample tube, it may not be possible to observe the cell during this transfer. Careful observation of the fluid height in a fine-bore pipet is usually sufficient to ensure the transfer of the cell and avoid adding an excessive volume of wash solution. The sample tube should be centrifuged briefly (a few seconds) to ensure that all liquid is at the bottom of the tube, and then returned to ice until the completion of all cell transfers.

### 3.1.3. Lysis Incubation

Lysis incubation should be carried out in a temperature-controlled block or thermal cycler separate from that used for amplification (*see* **Note 6**). Incubate samples at 50°C for 30 min, then 95°C for 15 min. It may be possible to shorten the 50°C incubation depending on the cell type. The high-temperature incubation is required to inactivate proteinase K completely. A heated cover must be in place over the samples to prevent condensation. If condensation is present on the cap or sides of the tubes following this incubation, subsequent amplification efficiency may be reduced. Reaction tubes can be frozen at –20°C (constant-temperature freezer) or placed on ice for immediate use.

## 3.2. LATE-PCR

There are three main criteria for LATE-PCR design. First, the concentration-adjusted melting temperature of the limiting primer ($T_m{}^L$) at the start of the reaction must be at least as high as that of the excess primer. This is achieved by making the limiting primer either longer or higher in percentage of guanine and cytosine (G + C) relative to the excess primer. Second, the concentration-adjusted temperature of the excess primer ($T_m{}^X$) must be reasonably close to the melting temperature of the double-stranded amplicon ($T_m{}^A$) in order for that primer to compete successfully with the accumulating single-stranded product for hybridization to the target strand. Third, if real-time detection is utilized, the concentration of the limiting primer should be chosen such that the limiting primer is depleted approximately when the probe signal reaches the detection threshold, i.e., at the $C_T$ value of the reaction.

### 3.2.1. Designing of Limiting and Excess Primers

Primers originally designed for symmetric PCR can be modified to fit LATE-PCR criteria (usually by lengthening the primer chosen as limiting), or primers can be newly selected according to those criteria. In either case, primer design software should be used to evaluate internal stability characteristics and 3' dimer formation in the same manner as would be done for symmetric PCR primers. Computer software can also be helpful in selecting primers. Preferred primer software provides input for primer concentration and should calculate $T_m$ according to nearest-neighbor methods using accurate thermodynamic values *(9–11)*. Do not rely on primer $T_m$ calculations based on the earlier estimates of nearest-neighbor thermodynamic values by Breslauer *(12)*.

When selecting new primers for LATE-PCR, it is useful to scan the sequences neighboring the site to be probed (e.g., mutation or polymorphism) for a region with relatively high GC content. The initial choice of a limiting primer can be made from that region. It generally does not matter which DNA strand is chosen for the sequence of the limiting primer as long as the hybridization probe is later chosen from a sequence on the same strand. An initial evaluation of the region to be amplified also can provide an estimate of $T_m^A$, which will be needed to determine the required $T_m^X$ value.

The concentration of limiting primers should be about 50 n$M$ (1.25 pmol/25-µL reaction) when used in combination with molecular beacons labeled with FAM or TET. At that concentration, a limiting primer length of approx 24–32 nucleotides is needed to achieve $T_m^L$ in the vicinity of 65°C (*see* **Note 7**). Excess primer concentration is usually 1 or 2 µ$M$. Optimal amplification efficiency and specificity are achieved with $T_m^X$ about 5° below $T_m^L$ when the primer concentration ratio is in the 20:1 to 40:1 range *(6)*. Primer $T_m$ calculations are made using the nearest-neighbor formula *(13)*:

$$T_m = \frac{\Delta H}{\Delta S + R \ln(C/2)} + 12.5 \log[M] - 273.15$$

The thermodynamic values $\Delta H$ and $\Delta S$ are calculated according to Allawi and SantaLucia *(9)*. $R$ is the universal gas constant and $C$ is the initial concentration of the primer. The salt correction is that of SantaLucia et al. *(14)* using $[M]$ as the total molar concentration of monovalent cations, sodium, and potassium in the PCR buffer. The $T_m$ calculations can be made using the MELTING program available on the Internet site http://bioweb.pasteur.fr/seqanal/interfaces/melting.html.

Another consideration in designing primers for LATE-PCR is $T_m^A$. That value depends primarily on amplicon length and GC content. Short amplicons (about 100 nt) are preferred for gene expression analysis or diagnosis of a specific genetic allele. When multiple alleles are tested or sequencing information is desired, longer regions can be successfully amplified. We have been able to amplify a 650-nt segment of the p53 gene using LATE-PCR criteria and sequence that product without purification (unpublished data).

Even for short amplicons, $T_m^A$ does not vary significantly with concentration, because the helix growth steps dominate the helix initiation step, producing a pseudo-first-order equilibrium for which no concentration effect is observed *(12)*. Therefore, good estimates of $T_m^A$ are obtained using a "%GC" formula *(15)*:

$$T_m^A = 81.5 + 16.6 \log \frac{[M]}{1 + 0.7[M]} + 0.41(\%G + \%C) - 500 \, / \, \text{length}$$

The formulas do not include a factor for magnesium concentration, which can raise the actual $T_m$ several degrees, but still provide valuable comparisons for designing amplification reactions.

As already mentioned, $T_m^X$ must be reasonably close to $T_m^A$ in order for that primer to compete successfully with the accumulating single-stranded product for hybridization to the target strand. We have observed the strongest real-time detection signals when $(T_m^A - T_m^X)$ is about 10 to 15°C, as calculated using these formulas *(6)*. Signal strength was lower as $(T_m^A - T_m^X)$ increased and was unacceptably low when it exceeded 20°C. Therefore, primers must have higher $T_m$ for amplicons that are long or GC rich.

### 3.2.2. Probe Design

We describe the design of molecular beacons for LATE-PCR, although it should be recognized that many other types of oligonucleotide probes can be used with this amplification technique (*see* **Note 8**). Molecular beacons are fluorescently labeled oligonucleotides that assume a stem-loop structure in the absence of homologous target, bringing a fluorophore on the 5' end of the molecule into close proximity of a quenching moiety (e.g., DABCYL) on the 3' end (**Fig. 1**) *(1)*. The molecular beacon is able to hybridize with a DNA strand (such as a PCR product) with sequence homologous to its loop. In that configuration, the fluorophore emits its fluorescent signal when illuminated at particular wavelengths. Thus, increasing PCR product in the presence of the homologous molecular beacon generates corresponding increases in fluorescent signal (**Fig. 2**). Multiple targets can be monitored in the same reaction by labeling different molecular beacon sequences with different fluorophores. LATE-PCR makes it possible to use molecular beacons with shortened loop sequences and greater allele discrimination.

The sequence of the molecular beacon loop (or any other probe that fluoresces on hybridization) must be chosen from the same DNA strand as the limiting primer. If the probe is used for distinguishing a single nucleotide polymorphism (SNP), that site should be in the center third of the loop. The $T_m$ of the beacon loop sequence to its target ($T_m^P$) should be at least 5° and preferably about 10° below $T_m^L$. This contrasts with the situation in symmetric PCR, in which $T_m^P$ must be greater than the primer $T_m$. The lower $T_m^P$ value ensures that the probe will not interfere with extension of the limiting primer. Thus, amplification efficiency during the exponential phase of LATE-PCR remains high even in the presence of high concentrations of probe.

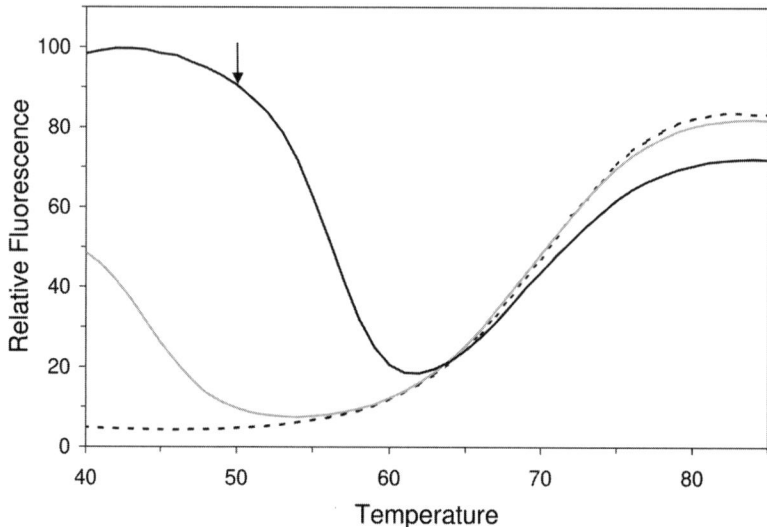

Fig. 3. Melting analysis of ΔF508 molecular beacon in absence of target (broken line), with mismatched normal allele target (gray line), and with ΔF508 target (black line). The measured $T_m$ of this molecular beacon with the complementary ΔF508 target is about 56°C. These results show that detection during PCR at 50°C fluorescence will provide close to maximum fluorescence with that target (arrow), but only background fluorescence with the mismatched target.

Good estimates of $T_m{}^P$ can be obtained using the same nearest-neighbor formula used for determining $T_m{}^L$ and $T_m{}^X$, even though variations in the molecular beacon stem do affect the empirically determined $T_m$ of the beacon-target hybrid. Stem designs are similar to conventional design, typically 5 to 6 bp, predominantly G and C. The stem $T_m$ is estimated using the intramolecular hybridization program mfold (16), available on-line at http://www.bioinfo. rpi.edu/applications/mfold/. mfold is also useful in identifying sequences that can form stable nonhairpin structures. Unlike conventional designs for molecular beacons in which $T_m{}^P$ and stem $T_m$ are both typically 7–10°C above the annealing temperature, we prefer to increase the stem $T_m$ 5–10°C above $T_m{}^P$ in order to ensure lower background fluorescence at the annealing temperature.

Molecular beacons should be tested using synthetic target oligonucleotides prior to use in LATE-PCR. The complementary oligonucleotide target should include at least 3 nt beyond each end of the molecular beacon loop, using the sequence of the target gene, so that possible interaction between the stem and target is included in the empirically determined $T_m$ (*see* **Note 9**). A melting analysis of molecular beacon in the absence of target, with complementary target, and with mismatched target (in the case of SNP analysis) is carried out to determine the best temperature for allele-specific detection (**Fig. 3**). Molecular beacon is used at a concentration that will be present during LATE-PCR, typically 1 μ*M*, and the concentration of targets should be about 0.5 μ*M*, the estimated final concentration of single-stranded product following LATE-PCR. Sodium, potassium, and magnesium concentrations should be the same as those used for amplification. Additional details on the design, synthesis, and testing of molecular beacon are available in the scientific literature *(17)* and on the Internet site http://www.molecular-beacons.org.

### 3.2.3. Components of LATE-PCR

With the exception of the primer and probe concentrations, other components used in LATE-PCR samples are similar to those used in symmetric reactions for single cells. The use of a "hot-start" method to prevent mispriming prior to the first denaturation step is required. Several commercially available Taq polymerases are modified so that they become active only after the initial high-temperature incubation. We prefer to use Taq polymerase with antibodies, because the required denaturation step is usually shorter. Taq polymerases from different commercial sources are supplied with buffers containing sodium Tris (or other buffer) and KCl. Begin testing using the recommended buffer solution, keeping in mind that varying the concentration of the monovalent cations will affect primer and probe $T_m$.

The dNTPs (specifically dATP, dCTP, dGTP, and dTTP) should be PCR grade and included at about 0.2 m*M* each. Higher concentrations may be needed for multiplex reactions, and lower concentrations are useful if the single-stranded product will be used directly for sequencing. Remember that dNTPs chelate magnesium ions and thereby affect the free magnesium concentration in the sample. Therefore, changes to dNTP concentration may affect reaction efficiency and specificity. We generally use a magnesium concentration of 3 m*M*. That concentration works well with most Taq polymerase enzymes and molecular beacon probes.

### 3.2.4. LATE-PCR Cycling Parameters

#### 3.2.4.1. INITIAL CYCLING STEPS AND DURATION

An initial denaturation step of 95°C for 2 min is followed by 25–35 initial cycles with steps for primer annealing, primer extension, and product denaturation. Fluorescence detection is not needed during these cycles. The annealing step should be no more than 10–15 s. Longer incubations promote nonspecific amplification. The extension step is usually carried out at 72°C, at which Taq polymerase has maximal activity. If amplicon size is only 100–200 nt, 15 s is more than sufficient to complete primer extension. A denaturation step of 5 s at 95°C should separate the DNA strands of most amplicons, enabling hybridization with primers during the subsequent annealing step.

Testing replicate samples containing low, equal concentrations of genomic DNA (e.g., 10 or 100 genome equivalents) at three or four different annealing temperatures is usually sufficient to identify optimal conditions (*see* **Note 10**). That optimum is usually close to the calculated $T_m^X$ value in samples containing 3 m$M$ magnesium, 20 m$M$ sodium, and 50 m$M$ potassium. Amplification efficiency at different annealing temperatures is evaluated by comparing mean $C_T$ values of replicate samples. Lower $C_T$ values (earlier detection) indicate higher amplification efficiency. Reaction specificity can be evaluated by analyzing products using gel electrophoresis. Alternatively, the DNA-binding dye SYBR Green I can be substituted for the hybridization probe, and product melting analysis can be done following amplification *(18)* (*see* **Notes 11** and **12**).

One of the difficulties of traditional asymmetric PCR is low amplification efficiency, which for real-time reactions causes delays in detection and the inability to obtain quantitative information. Another problem is the high level of nonspecific amplification, which can reduce the yield of specific product and the resulting signals from hybridization probes. By designing primers for which $T_m^L$ is higher than $T_m^X$, LATE-PCR makes it possible to use annealing temperatures that are low enough to ensure high amplification efficiency by the limiting primer, yet high enough to minimize mispriming by the excess primer.

### 3.2.4.2. FLUORESCENCE DETECTION DURING LINEAR AMPLIFICATION

Molecular beacon signals for single-copy targets usually reach detection threshold around cycle 40–45, depending on the detection equipment and the specific molecular beacon. Detection threshold will be reached about four cycles earlier for each 10-fold increase in the initial target concentration. Fluorescence detection should be included during cycling starting about 10 cycles before reaching threshold, and those initial values are then used to determine a fluorescence baseline for subsequent readings (*see* **Subheading 3.3**).

Detection can be carried out either during the annealing step of a standard thermal cycle or, preferably, during a step added after extension, since most of the amplicon strand detected by the probe remains single stranded during the linear phase of LATE-PCR. The temperature at which detection is done is chosen based on the tests with the probe and synthetic targets (*see* **Subheading 3.2.2.** and **Fig. 3**). That temperature should be low enough to provide strong signal from the complementary targets, but high enough to avoid signal from mismatched targets. Dropping the temperature a few degrees below the optimal annealing temperature at this point of the reaction usually does not present a problem in terms of nonspecific amplification. However, large drops in temperature should be avoided, because mispriming by amplicon strands may produce a phenomenon we refer to as "product evolution" (*see* **Note 12**).

The $C_T$ value of the reaction should be reached at or slightly before the limiting primer is depleted. Under these circumstances, the observed $C_T$ value will reflect the number of copies of the target sequence present in the sample at the start of the reaction, as in the case of symmetric real-time PCR. It may be necessary to test higher concentrations of the limiting primer (e.g., 100 or 200 n$M$) if the $C_T$ values are higher than anticipated or the subsequent rate of increase in fluorescence is low. Conversely, a nonlinear increase in fluorescence may reflect limiting primer concentrations that are too high. Remember that altering the concentration of the primers will change their $T_m$.

### 3.2.5. Preparation and Running of Diagnostic Assays on Lysed Cells

Large volumes of solutions containing all PCR reagents except Taq polymerase can be prepared and stored frozen in aliquots sufficient for concurrently tested samples, including positive and negative controls. Using the same mixture provides the highest reproducibility between assays run on different days. Taq polymerase should be added to the thawed aliquot just before use. The reagent solution should be thoroughly mixed before being added to individual samples containing lysed cells. A final volume of 25 µL is used for most applications. Samples should be kept on ice to ensure minimal polymerase activity during preparation. Even in the presence of antibodies, some mispriming may occur if samples are kept at room temperature for long periods prior to PCR.

The thermal cycler is programmed with the optimal cycling parameters with the detection step included about 10 cycles before the anticipated $C_T$ values. In most cases with single-copy genes, this will mean that the detection step will be added after the first 30 cycles. We typically run a total of 60 cycles. Specific requirements for selecting sample wells, programs, and detection wavelengths will vary with different thermal cyclers.

### 3.3. Analysis of Real-Time Signals

To display and analyze real-time signals properly, a baseline is set using fluorescence readings in the cycles before amplicon detection. Using a baseline normalizes background variations and gradual increases in fluorescence unrelated to the amplicon. The baseline readings can include any or all of the cycles before an increase in fluorescence. At least five cycles are usually necessary. Baselines are determined separately for each fluorophore used. Thermal cycler manufacturers typically suggest a threshold of 5 or 10 SDs above baseline detection values. We have found that choosing a threshold with a specific fluorescent value often provides better reproducibility between assays run at different times using the same PCR reagent mixture.

One of the advantages of real-time PCR is the ability to identify samples with atypical signal kinetics. Assay accuracy can be increased by excluding such samples from diagnosis *(19,20)*. LATE-PCR increases final fluorescence intensity and reduces sample-to-sample variation, thereby improving the kinetic analysis.

To establish limits for diagnosis, cells with known genotypes should be tested at the same time or at least using the same PCR mixture as unknown samples. Each genotype should be represented by about 10 or more samples. Mean values for $C_T$, increase in fluorescence (slope), and final fluorescence are determined for samples with the same genotype. Individual sample values are evaluated using the Extreme Studentized Deviate (ESD statistic) method to identify outliers *(21)*. Any positive control sample that does not yield the expected signals, or yields a value that is a statistical outlier, is not used for establishing diagnostic limits. We typically set those limits at 3 SDs from the means.

**Figure 4** illustrates this method for final fluorescence values obtained from human lymphoblasts homozygous or heterozygous for the ΔF508 mutation in the cystic fibrosis gene (*CFTR*), or homozygous normal at that locus. An initial assay was run to establish diagnostic limits, including the limits for final fluorescence indicated by the dashed lines for homozygous normal cells (box 1), homozygous mutant cells (box 2), or heterozygous cells (box 3). Because LATE-PCR yields a narrower range of final fluorescence values compared to other real-time methods, the size of these boxes is relatively small and gives a useful first step for data analysis. Individual data points shown include all results from a second assay using the same PCR reagent mixture, simulating "unknown" sample testing. The large majority of data points fall within the boxes established for the corresponding genotypes. A few samples did not give the expected results. Two samples with values outside the boxes might have been misdiagnosed but are excluded based solely on quantitative analysis of final fluorescence. First, one heterozygous cell yielded an extremely low fluorescence value for the ΔF508 allele (open diamond near the upper right corner of box 1). Such a preferential amplification result would have almost certainly been misdiagnosed as homozygous normal using conventional PCR and electrophoresis, but the molecular beacon provides the sensitivity to detect the mutant allele. Second, a single homozygous mutant sample generated a low-level signal for the normal allele, presumably owing to contamination (solid square to the left of box 3). That signal, however, was outside the limits for final fluorescence, and well outside the limits for $C_T$ value (latter not shown in **Fig. 4**) and, therefore, would not be misdiagnosed as heterozygous.

Evaluating $C_T$ values and rates of increase in fluorescence (slope) provides additional means for reducing misdiagnosis. **Figure 5** plots those values for the samples that yielded signals only from the normal allele, i.e., those indicated by the data points in box 1 of **Fig. 4**, including two heterozygous cell samples that failed to generate a ΔF508 allele signal. The diagnostic limits for $C_T$ values and rates of increase in fluorescence were established using data from the prior assay, as described in the preceding paragraph. One of the heterozygous cell samples generated a $C_T$ value above those limits for diagnosis as homozygous

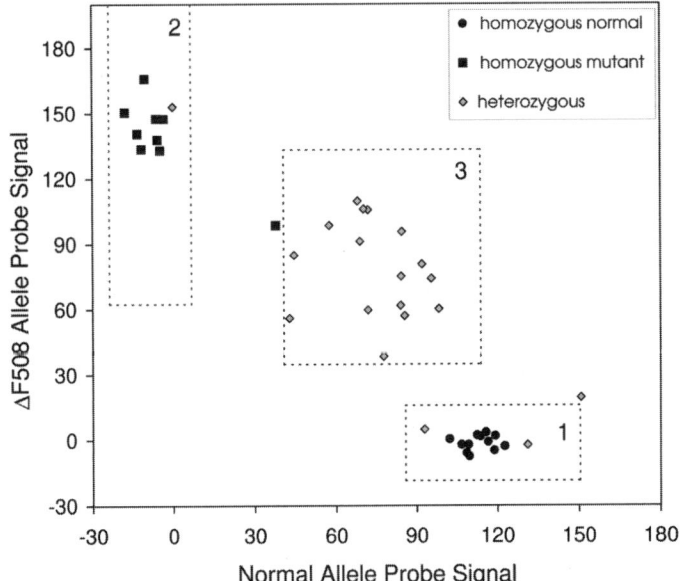

Fig. 4. Scatter plot of LATE-PCR final fluorescence values in replicate samples of individual lymphoblasts homozygous normal for *CFTR* (solid circles), homozygous for ΔF508 mutation (solid squares), or heterozygous for the ΔF508 mutation (shaded diamonds). The boxes labeled 1, 2, and 3 indicate diagnostic limits for those genotypes that were established by previously tested samples. The accuracy of the assay is improved by excluding from diagnosis all samples outside these limits. See the text for details.

normal. The other heterozygous cell sample gave values within the limits, although the slope was higher than that from any of the homozygous cells. In the setting of preimplantation genetic diagnosis (PGD) in which both parents carry the same mutant allele, failure to identify allele dropout (ADO) of the ΔF508 allele has no phenotypic consequence but becomes extremely important as the assay is extended to multiple mutation sites within a gene (*see* **Subheading 3.4.**). Analysis of the real-time signals using the 3-SD limits reduces misdiagnosis of ADO by about half for both symmetric PCR and LATE-PCR (*[7,20]*, unpublished data).

### 3.4. Assays for Detecting Compound Heterozygotes

Genotyping multiple mutation or SNP sites within a gene requires either a single amplicon that encompasses both sites or coamplification of the two regions using separate pairs of limiting and excess primers. If possible, the single amplicon approach is preferred, because it is simpler to design and optimize and provides a means to detect ADO in PGD cases. (Absence of the normal

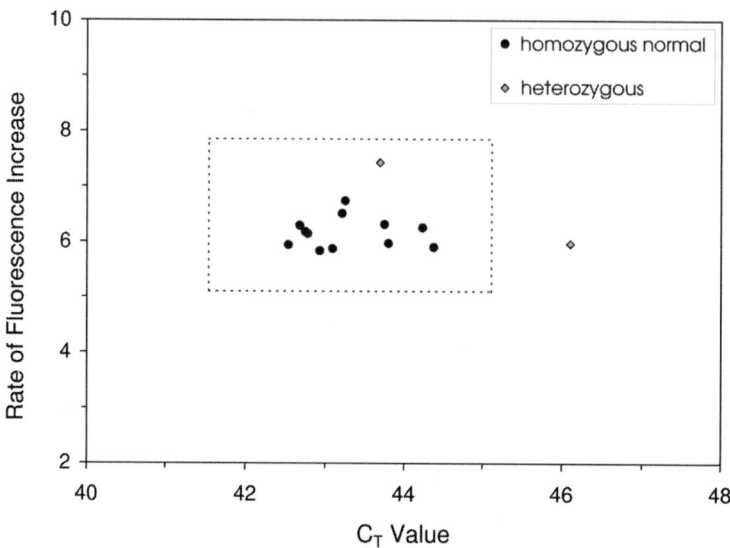

Fig. 5. Scatter plot of LATE-PCR $C_T$ values and rates of increase in fluorescence for samples in box 1 of **Fig. 4** (i.e., those generating only normal allele signal). Assay accuracy can be increased using the diagnostic limits indicated by the broken lines. A similar analysis can be done for samples generating 1 only ΔF508 allele signals (box 2 of **Fig. 4**) and for samples generating both signals (box 3 of **Fig. 4**). See the text for additional details.

allele signal at either site is an indication of ADO.) In contrast to symmetric real-time assays in which signal strength declines rapidly as amplicon size increases, LATE-PCR can generate strong signals with amplicons several hundred nucleotides long. Although we have not tested the limits in this area, strong signals have been obtained for a 650-nt amplicon. Limits may depend more on $T_m^A$ values, for reasons discussed in **Subheading 3.2.**, and thus will be longer when the %GC of the amplicon is low and relatively short when %GC is high.

In cases in which the distance between the mutation sites is too great for single amplicon design, the individual sites can be coamplified using LATE-PCR design criteria for each. Primers need to be evaluated for possible 3' dimer formation, as would be the case for any multiplex PCR, with particular attention paid to possible interaction between the two excess primers. Optimizing PCR reagent concentrations and the annealing temperature enables coamplification of both targets to similar levels. Moderate increases in the concentrations of one pair of primers can be used to equalize amplification efficiencies. We have successfully coamplified *CFTR* exon 10 and exon 11 sequences using LATE-PCR (unpublished data).

ADO has greater consequences when multiplex amplification is used in PGD to identify compound heterozygotes, because a failure to amplify either of the mutant alleles can lead to misdiagnosis and transfer of an affected embryo. Several PGD centers have implemented tests that include amplification of polymorphic sites found near the tested gene *(22)*. This approach, however, only detects loss of the entire region of the gene (e.g., owing to aneuploidy) or poor accessibility of the DNA in that entire region owing to inadequate cell lysis. Those situations are also detected by the absence of one of the normal allele sequences when testing two sites within the gene. When the target gene copies are both present, amplification from different sites within or neighboring the gene are independent events and, therefore, coamplification of neighboring polymorphic sites has limited value. By contrast, the analysis of real-time kinetics provides a means of identifying samples that exhibit atypical amplification including ADO. The strong signals and reliability of LATE-PCR offer the best opportunity to increase the diagnostic accuracy of single cell PCR.

## 4. Notes

1. PVP at 0.01 mg/mL in the final cell wash step in a cystic fibrosis assay does not delay detection *(19)*. Preliminary tests did show that detection was delayed slightly when higher concentrations of that solution were added to the lysis solution. Much larger and more variable detection delay was found with polyvinyl alcohol, possibly owing to interference with fluorescence detection.
2. We have observed that calcium can inhibit amplification. Serum and other culture additives such as hemoglobin, immunoglobulin, and heparin also interfere with amplification *(23,24)*. Adding BSA (nonacetylated, nuclease free) can improve PCR efficiency in the presence of some of these inhibitors *(24)*.
3. The volume of the lysis solution can be adjusted for the specific application and final PCR volume. Volumes below 10 µL can be used if the volume of transferred PBS is less than about 10% of that volume. Higher volumes are limited only by the volume of subsequently added PCR reagents and the final PCR volume.
4. A cell inadvertently introduced at any step can provide a DNA template for amplification. Contamination control measures should include dedicated pipettors for preparation of solution, aerosol-resistant pipet tips, lab coats, disposable caps, masks, extended-cuff gloves, and containment hoods, all in rooms separate from the PCR amplification area. Although ultraviolet treatment offers some protection from contaminating DNA, its effectiveness is limited. Treating surfaces with 10% bleach (1% sodium hypochlorite) is more effective for eliminating contaminating cells and DNA. The work area should be a "DNA-free zone" that resembles a "sterile field" in an operating room. Only clean gloves should touch items within that zone (e.g., sample tubes, pipettors, pipet tip boxes, reagent vials), and gloves that come in contact with any area outside the zone should be changed immediately. These procedures should be used when handling sample tubes at any step prior to PCR.
5. We have found that proteinase K–based lysis is not as effective in PCR tubes with glasslike properties, such as those for the Cepheid Smart Cycler, presumably owing to the adhesive properties of those tubes. Doing the lysis step in a standard sample

tube, then adding PCR reagents and transferring to the required PCR tube is an alternative, but losing material during that transfer increases the possibility of ADO. Preliminary results indicate that alkaline lysis *without* dithiothreitol (DTT) can produce acceptable results in the Smart Cycler tubes. DTT was present in the initial protocol for reducing protamines in sperm *(25)* but is unnecessary for most cell types, and residual DTT reduces PCR efficiency *(8)*.

6. Contamination of a thermal cycler block with PCR product is nearly impossible to avoid, even when great care is taken not to open tubes following PCR. If the same thermal cycle is used in a subsequent assay for the lysis incubation, it is possible to introduce those product molecules into samples when the tubes are opened to add PCR reagents. If the same block must be used for both lysis and PCR, it must be decontaminated between each assay. The block should be flooded with 10% bleach and then rinsed thoroughly with water.

7. If an adequate limiting primer cannot be designed using the actual DNA sequence of the target gene, $T_m^L$ can be increased by substituting one or two guanine bases for adenine near the 5' end of the primer. Hybridization of that primer with the initial target will have low-affinity G-T pairing, but not destabilizing mismatches, and subsequent hybridization with complementary amplicon strands will provide high amplification efficiency during the exponential phase of LATE-PCR. Another option is the addition of cytosine or guanidine to the 5' end of the primer, irrespective of the target sequence. Because the annealing temperature during the initial cycles cannot be lowered without risking mispriming by the excess primer, these options have obvious limits, particularly with a low initial target number, and, therefore, the $T_m$ of the limiting primer with the initial target sequence should not be more than 5°C below $T_m^X$.

8. We have recently developed allele-specific assays using double-stranded displacement probes *(26,27)*. These probes are easy to design and are relatively inexpensive, because each oligonucleotide is modified with a single fluorophore or quencher, not both. Extensive purification is not necessary, greatly increasing the manufacturing yield relative to dual-labeled probes such as molecular beacons. In general, the LATE-PCR benefits of increased signal strength and allele specificity can be accrued using any probe that signals on hybridization. Although TaqMan probes could be designed to work with LATE-PCR amplification, the need for hydrolysis requires that those probes have high melting temperatures and hybridize with the extension products of the limiting primer, rather than the accumulating single-stranded product. Therefore, benefits with TaqMan probes are limited.

9. When designing the molecular beacon stem, it is worthwhile to check for complementarities with nucleotides in the target sequence. It is usually possible to modify the stem slightly to minimize hybridization between the stem and target. Alternatively, those hybridizations can be allowed but should be taken into account when predicting $T_m^P$.

10. Alternatively, annealing temperature can be held constant and magnesium concentration is varied to identify optimal annealing conditions. We have observed that increasing the magnesium concentration from 3.0 to 3.5 m$M$ has a similar effect to

lowering the annealing temperature 2°. Large changes in magnesium concentration, however, may affect Taq activity and change hybridization characteristics of molecular beacons and other probes. In addition, note that these tests can be done on genomic DNA rather than single cells, since the limiting primer becomes depleted once it makes sufficient product to reach the detection threshold, regardless of the initial target concentration. Using 600 pg of DNA (equivalent to about 100 genomes) will lower the $C_T$ value about eight cycles compared to single cells but does not change the subsequent linear signal kinetics *(6)*.

11. SYBR Green I binds to double-stranded DNA regardless of nucleotide sequence. Fluorescence therefore plateaus after the limiting primer is exhausted. Following PCR cycling, fluorescence is monitored as temperature is gradually increased. As PCR products denature, a large drop in fluorescence is observed. Multiple drops in fluorescence, usually evaluated as "melting peaks" on plots of temperature vs the rate of decrease in fluorescence, indicate the presence of nonspecific product. Specific reactions should have a single melting peak about 3–6° above the calculated $T_m{}^A$ value, depending on the magnesium concentration.

12. If LATE-PCR is continued for many linear cycles, a second rise in SYBR Green fluorescence may be observed. This corresponds to a phenomenon that we call "product evolution," which involves the single strands priming on one another with a resulting increase in product size and melting temperature. Product evolution usually can be avoided by limiting the number of linear cycles and minimizing the drop in temperature needed for probe detection. In rare cases, it may be necessary to modify the 5' end of the limiting primer, thereby changing the 3' end of the amplicon single strands, in order to avoid this type of mispriming.

## Acknowledgments

We thank Aquiles Sanchez, John Rice, Cristina Hartshorn, Arthur Reis, Kevin Soares, and Jesse Salk for their contributions to the development and testing of LATE-PCR. This work was funded by Brandeis University.

## References

1. Tyagi, S. and Kramer, F. R. (1996) Molecular beacons: probes that fluoresce upon hybridization. *Nat. Biotechnol.* **14,** 303–308.
2. Heid, C. A., Stevens, J., Livak, K. J., and Williams, P. M. (1996) Real Time Quantitative PCR. *Genome Res.* **6,** 986–994.
3. Kostrikis, L. G., Tyagi, S., Mhlanga, M. M., Ho, D. D., and Kramer, F. R. (1998) Spectral genotyping of human alleles. *Science* **279,** 1228, 1229.
4. Gyllensten, U. B. and Erlich, H. A. (1988) Generation of single-stranded DNA by the polymerase chain reaction and its application to direct sequencing of the HLA-DQA locus. *Proc. Natl. Acad. Sci. USA* **85,** 7652–7656.
5. Sanchez, J. A., Pierce, K. E., Rice, J. E., and Wangh, L. J. (2004) Linear-After-The-Exponential (LATE)-PCR: an advanced method of asymmetric PCR and its uses in quantitative real-time analysis. *Proc. Natl. Acad. Sci. USA* **101,** 1933–1938.

6. Pierce, K. E., Rice, J. E., Sanchez, J. A., and Wangh, L. J. (2005) LATE-PCR: primer design criteria for high yields of specific single-stranded DNA and improved real-time detection. *Proc. Natl. Acad. Sci. USA*, **102**, 8609–8614.

7. Pierce, K. E., Rice, J. E., Sanchez, J. A., and Wangh, L. J. (2003) Detection of cystic fibrosis alleles from single cells using molecular beacons and a novel method of asymmetric real-time PCR. *Mol. Hum. Reprod.* **9**, 815–820.

8. Pierce, K. E., Rice, J. E., Sanchez, J. A., and Wangh, L. J. (2002) QuantiLyse™: reliable DNA amplification from single cells. *BioTechniques* **32**, 1106–1111.

9. Allawi, H. T. and SantaLucia, J. (1997) Thermodynamics and NMR of internal G·T mismatches in DNA. *Biochemistry* **36**, 10,581–10,594.

10. SantaLucia, J. (1998) A unified view of polymer, dumbbell, and oligonucleotide DNA nearest-neighbor thermodynamics. *Proc. Natl. Acad. Sci. USA* **95**, 1460–1465.

11. Owczarzy, R., Vallone, P. M., Gallo, F. J., Paner, T. M., Lane, M. J., and Benight, A. S. (1998) Predicting sequence-dependent melting stability of short duplex DNA oligomers. *Biopolymers* **44**, 217–239.

12. Breslauer, K. J. (1986) Methods for obtaining thermodynamic data on oligonucleotide transitions, in *Thermodynamic Data for Biochemistry and Biotechnology* (Hinz H. ed.), Springer-Verlag, New York, pp. 402–427.

13. Le Novère, N. (2001) MELTING, computing the melting temperature of nucleic acid duplex. *Bioinformatics* **17**, 1226, 1227.

14. SantaLucia, J., Allawi, H. T., and Seneviratne, P. A. (1996) Improved nearest-neighbor parameters for predicting DNA duplex stability. *Biochemistry* **35**, 3555–3562.

15. Wetmur, J. G. (1991) DNA probes: applications of the principles of nucleic acid hybridization. *Crit. Rev. Biochem. Mol. Biol.* **26**, 227–259.

16. Zuker, M. (2003) Mfold web server for nucleic acid folding and hybridization prediction. Nucleic Acids Res. **31**, 3406–3415.

17. Marras, S. A. E., Kramer, F. R., and Tyagi, S. (2003) Genotyping single nucleotide polymorphisms with molecular beacons, in *Single Nucleotide Polymorphisms: Methods and Protocols*, vol. 212 (Kwok, P. Y., ed.), Humana Press, Totowa, NJ, pp. 111–128.

18. Ririe, K. M., Rasmussen, R. P., and Wittwer, C. T. (1997) Product differentiation by analysis of DNA melting curves during the polymerase chain reaction. *Anal. Biochem.* **245**, 154–160.

19. Pierce, K. E., Rice, J. E., Sanchez, J. A., Brenner, C., and Wangh, L. J. (2000) Real-time PCR using molecular beacons for accurate detection of the Y chromosome in single human blastomeres. *Mol. Hum. Reprod.* **6**, 1155–1164.

20. Rice, J. E., Sanchez, J. A., Pierce, K. E., and Wangh, L. J. (2002) Real-time PCR with molecular beacons provides a highly accurate assay for Tay-Sachs alleles in single cells. *Prenat. Diagn.* **22**, 1130–1134.

21. Rosner, B. (1995) *Fundamentals of Biostatistics,* Wadsworth Publishing, Belmont, CA, pp. 277–282.

22. Rechitsky, S., Verlinsky, O., Amet, T., Rechitsky, M., Kouliev, T., Strom, C., and Verlinsky, Y. (2001) Reliability of preimplantation diagnosis for single gene disorders. *Mol. Cell. Endocrinol.* **183**, S65–S68.

23. Al-Soud, W. A., Jonsson, L. J., and Radstrom, P. (2000) Identification and charac terization of immunoglobulin G in blood as a major inhibitor of diagnostic PCR. *J. Clin. Microbiol.* **38,** 345–350.

24. Al-Soud, W. A. and Radstrom, P. (2001) Purification and characterization of PCR-inhibitory components in blood cells. *J. Clin. Microbiol.* **39,** 485–493.

25. Cui, X. F., Li, H. H., Goradia, T. M., Lange, K., Kazazian, H. H. Jr, Galas, D., and Arnheim, N. (1989) Single-sperm typing: determination of genetic distance between the G gamma-globin and parathyroid hormone loci by using the polymerase chain reaction and allele-specific oligomers. *Proc. Natl. Acad. Sci. USA* **86,** 9389–9393.

26. Li, Q., Luan, G., Guo, Q., and Liang, J. (2002) A new class of homogeneous nucleic acid probes based on specific displacement hybridization. *Nucleic Acids Res.* **30,** e5.

27. Cheng, J., Zhang, Y., and Li, Q. (2004) Real-time PCR genotyping using displacing probes. *Nucleic Acids Res.* **32,** e61.

# 8

## Efficient Isothermal Amplification of the Entire Genome from Single Cells

**Karen V. Schowalter, Jolene Fredrickson, and Alan R. Thornhill**

### Summary

Preimplantation genetic diagnosis for single gene disorders is usually performed using polymerase chain reaction (PCR)–based methodologies modified for use in single cells. At present, single cell PCR tests require costly and time-consuming development and validation of highly sensitive amplification strategies to cover a growing number of mutations responsible for genetic disease. Whole-genome amplification (WGA) provides an opportunity to amplify the genome from a single blastomere to a level at which multiple tests can be performed on the same cell. Early WGA methods (primer extension preamplification and degenerate oligonucleotide-primed PCR) have not proved sufficiently accurate and reliable for routine clinical use. However, WGA using multiple displacement amplification (MDA) offers approx 5 million–fold amplification with fidelity, apparently sidestepping the limitation of a single cell, and is sufficient for use in most off-the-shelf molecular tests. This chapter describes an optimized MDA protocol for the preparation of genomic DNA from single fibroblasts.

**Key Words:** Whole-genome amplification; multiple displacement amplification; single cell polymerase chain reaction; preimplantation genetic diagnosis; quality assessment; cell lysis.

## 1. Introduction

Preimplantation genetic diagnosis (PGD) following in vitro fertilization is now well established clinically as an alternative to conventional prenatal diagnosis in couples at risk of having children with an inherited disease *(1)*. Tests for the identification of single gene disorders in eggs or preimplantation embryos are usually performed using polymerase chain reaction (PCR)–based methodologies modified for use in single cells—the characteristics and limitations of which have been described previously in great detail *(2)*. Moreover, the genetic conditions for which PGD has been applied are numerous and the various methods used for diagnosis reflect the heterogeneity of causative mutations (*see* **ref. 2** for a review). However, the number of specific single cell tests available today represents only a

From: *Methods in Molecular Medicine: Single Cell Diagnostics: Methods and Protocols*
Edited by: A. R. Thornhill © Humana Press Inc., Totowa, NJ

small percentage of molecular diagnostic tests offered worldwide for more routine testing such as carrier screening or prenatal diagnosis.

The main challenge to the widespread implementation of PGD for single gene disorders is the provision of sufficient resources in order to offer a comprehensive menu of single cell–sensitive tests covering a wide range of inherited disorders and the many mutations responsible for those disorders. At present, single cell PCR tests require costly and time-consuming development and validation of highly sensitive amplification strategies that are highly susceptible to errors through contamination with foreign or previously amplified DNA and allele dropout (ADO). Hence, isolated clean-room facilities and stringent precautions are essential throughout to avoid misdiagnoses, thereby excluding the participation of molecular diagnostic laboratories where amplification of similar sequences from large numbers of samples is routine and PCR products are handled openly on the laboratory bench.

By contrast, the widespread application of preimplantation aneuploidy screening involves a single test using a generic fluorescence *in situ* hybridization (FISH) method that can be learned quickly and the results scored relatively easily. Using FISH, the perils of contamination are far less problematic. Furthermore, aside from the micromanipulation apparatus, no special equipment or laboratory conditions are required for efficient cell preparation.

Multiplexing, or the simultaneous detection of multiple specific sequences in a single PCR assay on a single cell, goes some way toward offering a single test to cover several mutations directly or indirectly (by linkage) of mutations although extensive optimization is required and preferential amplification and ADO still occur at appreciable rates *(3,4)*. Furthermore, a single multiplexing reaction still represents a one-shot approach in which stochastic PCR failure on that occasion prevents any diagnosis from that cell.

An exciting new prospect for single cell molecular analysis has been the development of isothermal whole-genome amplification (iWGA). From a starting template of 7 pg of DNA in a single cell, it is possible to yield 30 μg of DNA using iWGA (~5 million–fold amplification), apparently sidestepping the limitation of a single cell and sufficient for use in most off-the-shelf molecular tests. Previous WGA methods such as primer extension preamplification (PEP) and degenerate oligonucleotide primed PCR (DOP-PCR) are effective at amplifying most of the genome from a single blastomere *(5,6)*. However, the drawback with PEP is that it amplifies only between 30 and 100 times the starting template. Degenerate oligonucleotide primed PCR (DOP-PCR) amplifies a similar proportion of the genome to PEP *(7)*, but to a much more significant level, with a single cell providing enough DNA for more than 100 subsequent PCR amplifications. Furthermore, sufficient DNA is produced to allow additional experimental procedures such as comparative genomic hybridization for the detection of chromosome copy number *(8)*. Both PEP and DOP have been used for clinical PGD, but a significant drawback of both methods is that amplification of repetitive DNA sequences, such as short tandem repeats, is error prone if performed on WGA products. In some studies, >50% of fragments amplified differed from their expected size *(8,9)*. Such errors would currently rule out the use of WGA for the clinical diagnosis of triplet repeat expansion diseases or diagnoses based on linkage by microsatellite analysis. ADO rates after PEP and DOP-PCR are comparable with those obtained by direct amplification of single cell loci.

The ability to test for more than one gene sequence in a single cell as part of clinical PGD will allow the following: a test for the specific mutation (if known), confirmation of the diagnosis (internal quality control), external quality control (sending a portion of the sample to other laboratories), and additional assays including linked and unlinked markers to identify ADO and contamination. Further information would include chromosomal aneuploidy assessment and human leukocyte antigen typing. Moreover, WGA provides a supply of sample DNA that can be reassessed, allowing confirmation of diagnosis using the same or different methods or the analysis of other genes or even chromosomal status.

Recently, WGA using the bacteriophage f29 DNA polymerase for isothermal multiple displacement amplification (MDA) has been reported *(10,11)*. The f29 DNA polymerase has high processivity, generating amplified fragments of 10 kb by strand displacement, and has proofreading activity resulting in lower misincorporation rates compared with Taq polymerase. The random hexamer primers must be thiophosphate modified to protect them from degradation owing to 3' –5' exonuclease proofreading activity of the f29 DNA polymerase *(12)*. iWGA directly from clinical samples such as blood and buccal swabs has allowed high-throughput genotyping without the need for time-consuming DNA purification steps *(13,14)*. Sequence representation in the amplified DNA assessed by multiple single-nucleotide polymorphism analysis is equivalent to genomic DNA when amplifying from as little as 0.3 ng of target DNA (or ~50 single cell equivalents), and amplification bias is superior to that of PCR-based methods *(15)*.

Here we investigate the use of iWGA from single and small numbers of cells (in this case isolated fibroblasts), as a generic PGD assay to precede specific testing for any single gene disorder and chromosomal information. Several groups have used the iWGA kit described here, but with little in the way of modifications *(16–18)*. However, because the kit was not designed for single cells, we believe that modifications are necessary to achieve optimal results from single cells.

Preliminary experiments in our laboratory were designed to determine (1) optimal lysis conditions for single cells and low numbers of cells, (2) optimal polymerase enzyme concentrations, and (3) optimal reaction times to obtain the highest yield of DNA in combination with the most accurate results in a time frame suitable for use in clinical PGD. This chapter describes the optimized protocol for preparing genomic DNA from single fibroblasts.

## 2. Materials

### 2.1. Sources of Single Cells

Fibroblast cell line GM13591 with known mutation in the CFTR gene (purchased from Coriell Cell Repositories, Camden, NJ, http://www.coriell.org): Allele one carries a deletion of codon 508 (CTT) in exon 10, which leads to deletion of phenylalanine-508 [PHE508DEL], and allele two carries a G-to-A transition at nucleotide 482 (482G>A) in exon 4, resulting in a change of arginine to histidine [ARG117HIS (R117H)]. Analysis of a DNA variant in a noncoding region of the CFTR gene (polypyrimidine tract in intron 8) showed that this donor has alleles 5T/9T. The donor subject is homozygous for a C>G transversion at nucleotide 187 in exon 2 of the HFE (HLA-H) gene [187C>G], resulting in a substitution of aspartic acid for histidine at codon 63 [His63Asp (H63D)] (*see* **Note 1**).

### 2.2. Cell Culture

1. RPMI-1640 (Irvine Scientific).
2. L-Glutamine (29.2 mg/mL) (Irvine Scientific).
3. Fetal bovine serum (FBS) (Irvine Scientific).
4. Penicillin G (10,000 U/mL) (Irvine Scientific).
5. Streptomycin sulfate (10,000 µg/mL) (Irvine Scientific).
6. Incubator suitable for cell culture (5% $CO_2$).

### 2.3. Preparation of Single Cells

1. Phosphate-buffered saline (PBS)/polyvinylpyrrolidone (PVP) solution: PBS, pH 7.2, supplemented with 5 mg/mL of PVP. PVP is added to prevent cells from adhering to each other and to the bottom of the dish.
2. Handheld pipettor device for microcapillary tubes (e.g., Stripper; MidAtlantic Diagnostics)
3. Microcapillary tubes with an I.D. of 50 µm (e.g., PGD Stripper tip; MidAtlantic Diagnostics).
4. Thin-walled PCR tubes (0.5 mL) with flat caps (Applied Biosystems, Foster City, CA).
5. Falcon 1006 plastic Petri dish.
6. Stereomicroscope with large working distance and magnification range of H20 to H100 (e.g., Olympus CK2).

### 2.4. Lysis of Single Cells

1. Lysis buffer: 200 m*M* KOH (*see* **Note 2**).
2. Neutralization buffer: 300 m*M* KCl; 900 m*M* Tris-HCl, pH 8.3; 200 m*M* HCl.

## 2.5. iWGA (MDA) on Single Cells

1. Repli-g (Qiagen, Germantown, MD) Kit components:
   a. Vial of 4X Mix (0.32 mL).
   b. DNA Polymerase Mix (12.5 μL).
   c. Control gDNA template (25 μL) (10 ng/μL).
   d. Solution B (5 mL) (Lysis Stop Buffer = Neutralization buffer; *see* **Subheading 2.4., item 2**).
   e. 1X PBS (5 mL) (pH 7.5).
   f. Dithiothreitol (DTT) (1 mL).
   g. Empty vial for Solution A (Alkaline Solution), prepared fresh.

## 2.6. Quantification of Amplified Product

1. PicoGreen dsDNA Quantation Kit (P-7589; Molecular Probes, Invitrogen).
2. PicoGreen dsDNA Quantitation Reagent (Molecular Probes, Invitrogen).
3. 20X Tris-EDTA (TE) stock buffer.
4. Lambda DNA standard (100 μg/mL) (Invitrogen).

## 2.7. Quality Assessment of Amplified Product Using DNA Fingerprint

All of the following reagents and consumables are from Applied Biosystems.

1. ABI 3100 Pop-4 Polymer.
2. ABI Prism Linkage Mapping Set (12 chromosomes).
3. ABI Genotyper software.
4. ABI 3100 capillary array.
5. ABI 3100 10X buffer.
6. Hi-Di Formamide.
7. ABI 3100 GS-400HD (Rox) size standard.
8. MicroAmp Optical 96-well reaction plates.
9. 96-Well plate septa.

# 3. Methods

## 3.1. Preparation of Fibroblast Suspensions

Fibroblast cell lines with known mutations are normally shipped in small flasks (frozen on dry ice) containing medium with only 5% FBS and no glutamine, to slow down cell proliferation during transportation.

1. Place newly received flasks in an incubator at 37°C overnight without opening to allow the cells to settle.
2. Discard the shipping medium from the flasks and replace with freshly prepared RPMI-1640 working solution (add 50 mL of FBS, 6 mL of penicillin, streptomycin, and L-glutamine solution to a 500-mL bottle).
3. Once the cell line is approx 70% confluent (yellow, cloudy medium), remove the cell line to a laminar flow hood.
4. Using a new sterile 10-mL pipet, pipet the medium into a 15-mL labeled centrifuge tube.

5. Feed a T75 flask with 10 or 20 mL of complete RPMI-1640 depending on the amount of supernatant removed.
6. Centrifuge (Beckman TJ-6) at 1700 rpm for 8 min, aspirate off the medium to 0.5 mL, and discard.
7. Resuspend the pellet with 3 mL of PBS/PVP. The cells are now ready for single cell sorting.

### 3.2. Single Cell Sorting (see Fig. 1)

Prepare three 50-µL drops and several rows of 5-µL drops of PBS/PVP in a Falcon 1006 dish to accommodate the number of cells to be prepared using a single small drop for each cell.

1. Add 10 µL of the cell suspension (from **step 7** in **Subheading 3.1.**) to a 35-mm-diameter Petri dish containing 3.0 mL of PBS/PVP.
2. Transfer approx 100 cells from the dish to the first 50-µL drop using a stripper or equivalent handheld device (*see* **Note 3**) with a maximum volume of 3 µL.
3. Transfer between 15 and 20 cells from the first PBS/PVP drop into drops two and three for rinsing in several locations within the large drops. Thereafter transfer single cells into the smaller drops (one time use only) before transferring to a microcentrifuge tube.
4. Aspirate one cell from the small PBS/PVP drop and expel with ~0.5 µL of fluid into the bottom of a 0.5-mL sterilized PCR tube preloaded with 2.5 µL of PBS under microscopic visualization. If not visualized, rinse the tip of the pipet in a clean drop to ensure that the cell was expelled.
5. Isolate more single cells separately from the PBS/PVP drops using the same stripper tip.
6. Create wash drop (negative) controls by aspirating a small amount of PBS/PVP from the last drop from which a cell was taken, and transfer into a 0.5-mL sterilized microcentrifuge tube (*see* **Note 4**).
7. Maintain the tubes at room temperature until iWGA (**Subheading 3.4.**).

### 3.3. Lysis of Single Cells

1. Add 3.5 µL of lysis buffer to the 0.5-mL PCR tube containing a single cell. When preparing lysis buffer from Solution A, the kit instructions suggest a 1:10 dilution with 1 *M* DTT. For our modification, instead of DTT, our 1:10 dilution was prepared with sterile water (e.g., 3.5 µL of water into 31.5 µL of solution A to prepare 35 µL for 10 reactions). Incubate the tube on ice for 10 min.
2. Add 3.5 µL of neutralization solution (Solution B) to the tube to stop the reaction.

### 3.4. iWGA (MDA) on Single Cells

1. To the tube containing the lysed and neutralized single cell solution (10 µL), add 40 µL of the master mix (follow the manufacturer's protocol exactly, i.e., 27 µL of nuclease-free water, 12.5 µL of 4X buffer, and 0.5 µL of Phi polymerase enzyme; *see* **Note 5**).

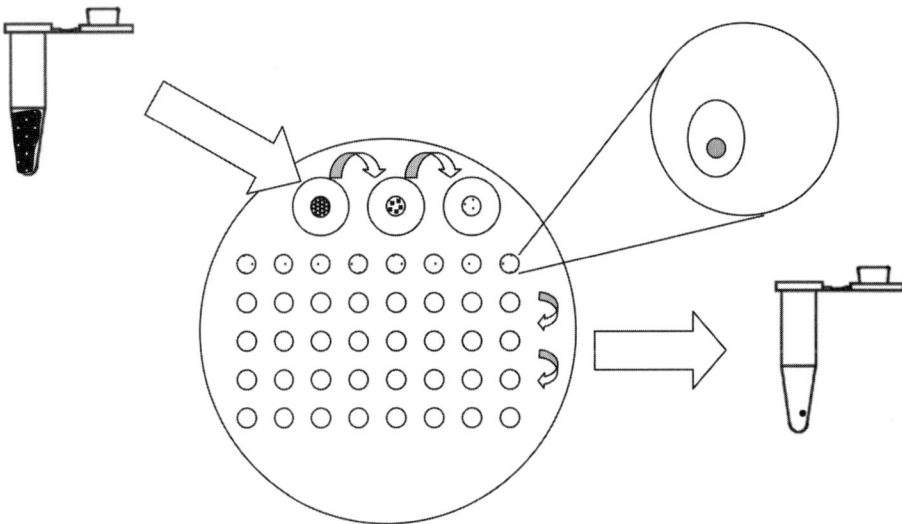

Fig. 1. Single cell sorting procedure. A cell suspension is diluted through drops of PBS/PVA placed in a Petri dish to the point where there are only three to five cells in PBS. A single cell is washed several times through different droplets and then transferred into a 0.5-mL PCR tube. PBS is aspirated from the final drop and transferred into a 0.5-mL PCR tube for use as a negative "wash-drop blank" control.

2. Briefly centrifuge the mixture , place in a Minicycler Thermal Cycler (MJ Research), and incubate at 37°C for 6 h (*see* **Note 6**).
3. Store the amplified product at between 4 and –80°C (depending on the length of time required).

## 3.5. Quantification of Amplified Product Using PicoGreen (19)

1. Prepare a 2 µg/mL solution of lambda DNA standard and dilute to a DNA concentration of 0–500 ng/mL. Pipet in duplicates in a microtiter plate.
2. Prepare unknown samples by diluting 2 µL of sample with 98 µL of 1X TE buffer. Pipet in duplicates into the microtiter plate. Add 100 µL of a 1:200 dilution of Pico Green to each well of standard dilutions and unknowns.
3. Open the Xflour4 program from Genios and read the samples. The program automatically calculates the DNA concentrations.

## 3.6. Quality Assessment of Amplified Product Using DNA Fingerprint/ Microsatellite Marker Analysis (see Note 7)

ABI Prism Markers from the Linkage Mapping Set Version 2.5 are used for PCR.

1. Set up six separate multiplex PCR reactions with two markers in each (markers used are D7S484, D14S70, D13S158, D21S1252, mycl, D10S197, D16S520, D2S2368, D15S1002, D6S441, D17S250, D8S262).

2. Perform PCR according to the manufacturer's specifications.
3. After PCR, pool and dilute the products of all six reactions in formamide and run on the ABI 3100.
4. Use the Genotyper Software program for fragment size analysis (*see* **Fig. 2** for a typical example of analysis following iWGA on single fibroblasts).

### 3.7. Performing of a Diagnostic Assay Using MDA DNA

If the concentration of DNA (as measured by PicoGreen assessment) is high, dilution of the amplified product may be required prior to diagnostic use. For routine clinical molecular genetics testing, the standard concentration for most PCR assays is 100–200 ng/µL with 1 µL added per reaction. Similarly, PCR products may be diluted before running on the ABI 3100 Genetic Analyzer if signals are initially off the scale and become difficult to call or interpret accurately. Using the optimized method to perform iWGA from single fibroblasts described here, we analyzed up to 63 different loci from the amplified DNA using different diagnostic platforms (including fluorescent PCR, Luminex technology, ethidium gel-based PCR, and Light Cycler technology). Most of the tests were previously optimized using standard DNA concentrations (e.g., 250 ng) rather than single cells, and we initially used 0.5–1 µL of the iWGA product for each reaction (which was frequently too concentrated). Loci analyzed included the DF508 deletion causing cystic fibrosis, CF panel, Tag-It CFTR 40+4 loci (Luminex, TM Biosciences) polymorphic microsatellite sequences used in DNA fingerprinting (Fluorescent PCR with ABI Prism Masters from the Linkage Mapping Set version 2.5), and C282Y and H63D mutations in the HFE gene (Lightcycler; Roche). Of 1751 loci analyzed from single cells after iWGA, 1653 (94.4%) gave interpretable results with an overall allele dropout rate of 18.7% (range: 4–89%).

### 3.8. Generation of More Amplified DNA from Previously Amplified DNA

If the original DNA template is precious, it may be necessary to reamplify the amplified product after performing multiple assays to prevent it from running out. Reamplification of iWGA product is simple. One microliter of iWGA product from a single cell is added to 1.5 µL of TE solution before denaturing with 2.5 µL of Denaturing Solution (solution A) and incubating for 3 min at room temperature. After neutralizing with 5 µL of Solution B, 40 µL of Master mix (water, 4X mix, and enzyme) is added. There follows a 16-h incubation at 30°C, 65°C for 3 min before holding at 4°C. One microliter of the resultant WGA product was used for 58 of the 63 assays described in **Subheading 3.7.** The results using the reamplified product were highly concordant with results from the original amplified product (*see* **Fig. 3**).

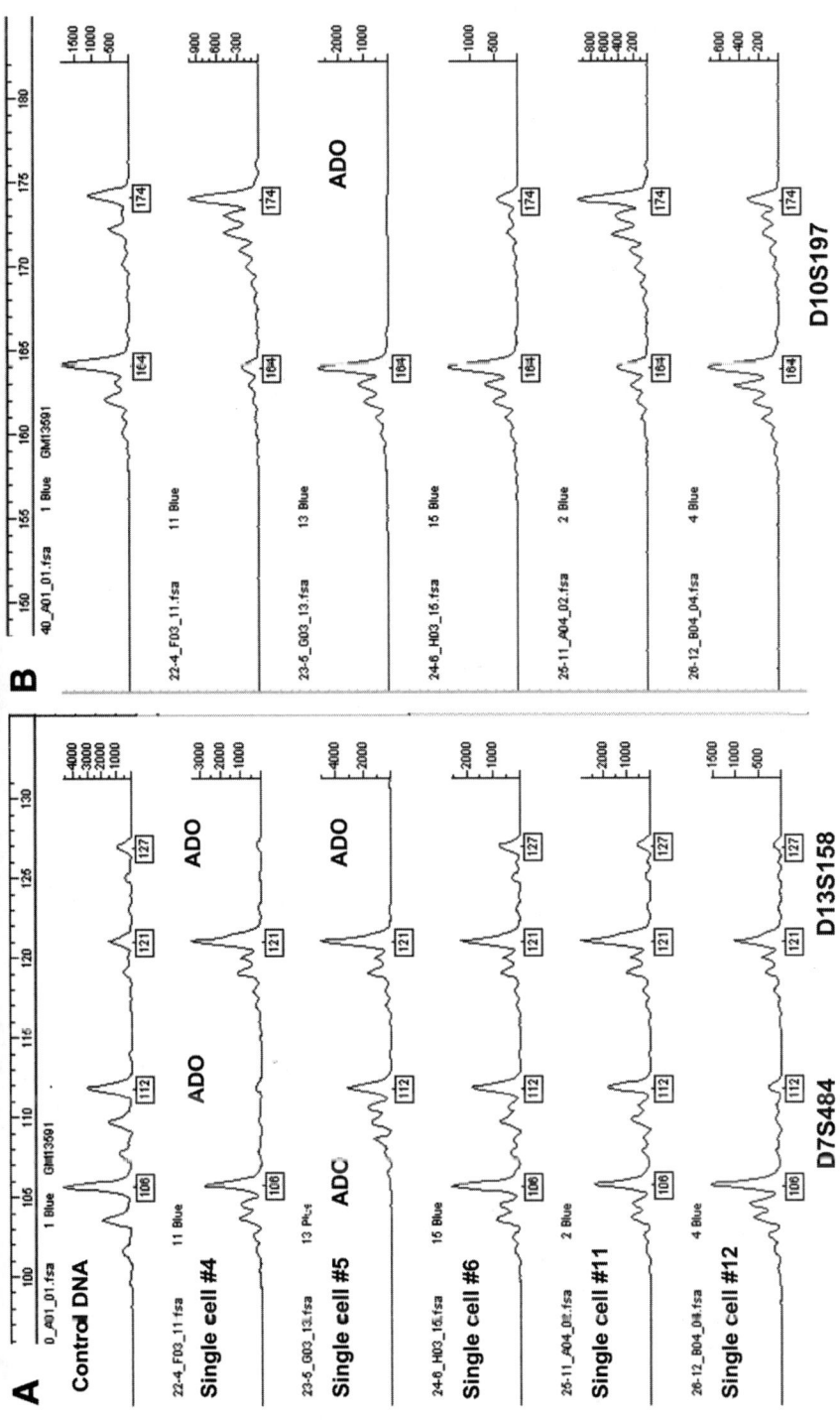

Fig. 2. Analysis of control DNA and five different single cells with various microsatellite markers following iWGA. (A) Markers for chromosome 7 (D7S484) and 13 (D13S158) are heterozygous. ADO is observed in cells 4 and 5. (B) Marker for chromosome 1C (D10S197) is heterozygous. ADO is observed in cell 5.

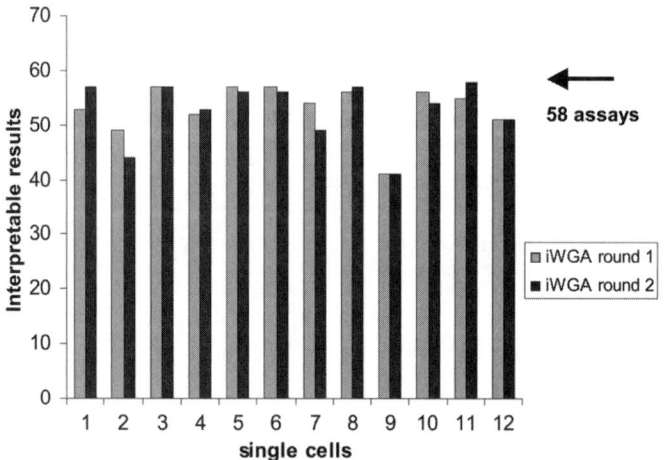

Fig. 3. Comparison between iWGA of single cells and reamplification of iWGA DNA product from corresponding single cell. A high degree of concordance with respect to proportions of interpretable results (using the exact same 58 assays) is observed following reamplification of the iWGA product. Overall, there are approximately the same number of "failures" from both the original iWGA product and its reamplified counterpart. Reamplification does not appear to compromise the integrity of the sample and the accuracy of the results obtained from it.

## 4. Notes

1. When developing single cell protocols, one should be aware that there is a cell-type effect on amplification and ADO rates. Lysis and amplification may be different in fibroblasts in comparison to lymphocytes, buccal cells, and blastomeres *(20)*, probably as a result of differential protein composition. As a consequence, protocols may need to be optimized for each specific cell type.
2. Assuming successful transfer of a high-quality nucleated cell, the cell lysis protocol used also influences amplification success. However, there is no consensus as to which lysis buffer is the most effective *(21)*. We assumed that the lysis buffer prior to WGA would be at least as critical to success as when direct PCR is applied to a single cell. Using KOH only as lysis buffer was previously shown to be slightly more effective than KOH + DTT on primary lymphocytes *(22)*, presumably because DTT can inhibit PCR. The DTT present in many lysis protocols was originally included to break disulfide bonds in condensed sperm heads, and this rigorous degree of lysis may be superfluous for other cell types. In our preliminary optimization experiments, KOH alone gave the best results for isothermal amplification and subsequent analysis.
3. Mouth pipetting is prohibited by most health and safety agencies. We have found that, after a period of learning, handheld pipetting using commercially available plastic stripper tips in conjunction with a stripper device can be used effectively to handle single cells.

4. The purpose and effectiveness of the wash-drop control is controversial. The number of blanks to include in assay development and clinical cases presents something of a dilemma in view of the calculation that 300 negative (reaction mix only) blanks are required to ensure that the contamination rate is <1% *(23)*. A two-stage testing procedure has been suggested to maintain this low contamination rate. Prior to clinical implementation, a large series of blanks (e.g., 100) should be run. Afterward, smaller series should be run periodically *(23)*.

5. When optimizing PCR amplifications for low-concentration DNA (e.g., extractions from paraffin-embedded material), increasing the amount of Taq polymerase can help achieve amplification of the target. For this reason, we used both the recommended concentration of enzyme (0.5 µL per the manufacturer's instructions) and double this amount (i.e., 1 µL per 50-µL reaction). We detected no significant differences between these two conditions.

6. According to the manufacturer, the iWGA enzymatic reaction plateaus after 6 h. Since 6 h (vs 16 h, the suggested time in the manufacturer's protocol) is a more realistic and convenient time frame in which to perform clinical PGD, we compared 6- and 16-h incubation times. Preliminary experiments yielded equivalent results, although we speculate that with low template availability, ADO could be exacerbated with extended reaction times, but this has not been formally tested.

7. Quality assessment prior to diagnostic use is a critical step. The Repli-g Kit developers used to provide a genotyping service that provided a crude quality assessment, but this was not optimized for single cells. Instead, we used a multiplex assay covering microsatellite markers on 12 different chromosomes (used for DNA fingerprinting; *see* **Subheading 3.6.**). If high fidelity and amplification efficiency was observed for this panel for a specific single cell, this was generally true for the other assays.

## Acknowledgments

We are grateful to David Walker for help with single cell preparation, Rick Helgemo for fibroblast culture and clinical technologists from the Mayo Clinic Molecular Genetics testing laboratory for running hundreds of WGA samples in parallel with clinical samples.

## References

1. Verlinsky, Y., Cohen, J., Munne, S., Gianaroli, L., Simpson, J. L., Ferraretti, A. P., and Kuliev, A. (2004) Over a decade of experience with preimplantation genetic diagnosis: a multicenter report. *Fertil. Steril.* **82(2)**, 292–294.
2. Thornhill, A. R. and Snow, K. (2002) Molecular diagnostics in preimplantation genetic diagnosis. *J. Mol. Diagn.* **4,**11–29.
3. Findlay, I., Matthews, P. L., Mulcahy, B. K., et al. (2001) Using MF-PCR to diagnose multiple disorders from single cells: implications for PGD. *Mol. Cell. Endocrinol.* **183(Suppl. 1),** S5–S12.
4. Katz, M. G., Mansfield, J., Gras, L., Trounson, A. O., and Cram, D. S. (2002) Diagnosis of trisomy 21 in preimplantation embryos by single-cell DNA fingerprinting. *Reprod. Biomed. Online* **4(1),** 43–50.

5. Snabes, M. C., Chong, S. S., Subramanian, S. B., et al. (1994) Preimplantation single-cell analysis of multiple genetic loci by whole-genome amplification. *Proc. Natl. Acad. Sci. USA* **91,** 6181–6185.

6. Sermon, K., Lissens, W., Joris, H., Van Steirteghem, A., and Liebaers, I. (1996) Adaptation of the primer extension preamplification (PEP) reaction for preimplantation diagnosis: single blastomere analysis using short PEP protocols. *Mol. Hum. Reprod.* **2(3),** 209–212.

7. Telenius, H., Carter, N. P., Bebb, C. E., Nordenskjold, M., Ponder, B. A., and Tunnacliffe, A. (1992) Degenerate oligonucleotide-primed PCR: general amplification of target DNA by a single degenerate primer. *Genomics* **13(3),** 718–725.

8. Wells, D., Sherlock, J. K., Handyside, A. H., et al. (1999) Detailed chromosomal and molecular genetic analysis of single cells by whole genome amplification and comparative genomic hybridisation. *Nucleic Acids Res.* **27,** 1214–1218.

9. Foucault, F., Praz, F., Jaulin, C., and Amor-Gueret, M. (1996) Experimental limits of PCR analysis of (CA)n repeat alterations. *Trends Genet.* **12(11),** 450–452.

10. Dean, F. B., Hosono, S., Fang, L., et al. (2002) Comprehensive human genome amplification using multiple displacement amplification. *Proc. Natl. Acad. Sci. USA* **99,** 5261–5266.

11. Lasken, R. S. and Egholm, M. (2003) Whole genome amplification: abundant supplies of DNA from precious samples or clinical specimens. *Trends Biotechnol.* **21,** 531–535.

12. Dean, F. B., Nelson, J. R., Giesler, T. L., et al. (2001) Rapid amplification of plasmid and phage DNA using Phi 29 DNA polymerase and multiply-primed rolling circle amplification. *Genome Res.* **11,** 1095–1099.

13. Hosono, S., Faruqi, A. F., Dean, F. B., et al. (2003) Unbiased whole-genome amplification directly from clinical samples. *Genome Res.* **13,** 954–964.

14. Mai, M., Hoyer, J. D., and McClure, R. F. (2004) Use of multiple displacement amplification to amplify genomic DNA before sequencing of the alpha and beta haemoglobin genes. *J. Clin. Pathol.* **57(6),** 637–640.

15. Lovmar, L., Fredriksson, M., Liljedahl, U., et al. (2003) Quantitative evaluation by minisequencing and microarrays reveals accurate multiplexed SNP genotyping of whole genome amplified DNA. *Nucleic Acids Res.* **31(21),** e129.

16. Handyside, A. H., Robinson, M. D., Simpson, R. J., Omar, M. B., Shaw, M. A., Grudzinskas, J. G., and Rutherford, A. (2004) Isothermal whole genome amplification from single and small numbers of cells: a new era for preimplantation genetic diagnosis of inherited disease. *Mol. Hum. Reprod.* **10(10),** 767–772.

17. Hellani, A., Coskun, S., Benkhalifa, M., Tbakhi, A., Sakati, N., Al-Odaib, A., and Ozand, P. (2004) Multiple displacement amplification on single cell and possible PGD applications. *Mol. Hum. Reprod.* **10(11),** 847–852.

18. Hellani, A., Coskun, S., Tbakhi, A., and Al-Hassan, S. (2005) Clinical application of multiple displacement amplification in preimplantation genetic diagnosis. *Reprod. Biomed. Online* **10(3),** 376–380.

19. Enger, U. (1996) Use of the fluorescent dye PicoGreen™ for quantification of PCR products after agarose gel electrophoresis. *BioTechniques* **21(3),** 372–374.

20. Rechitsky, S., Strom, C., Verlinsky, O., et al. (1998) Allele dropout in polar bodies and blastomeres. *J. Assist. Reprod. Genet.* **15(5),** 253–257.
21. Geraedts, J., Handyside, A., Harper, J., et al. (1999) ESHRE Preimplantation Genetic Diagnosis (PGD) Consortium: preliminary assessment of data from January 1997 to September 1998. *Hum. Reprod.* **14(12),** 3138–3148.
22. Pierce, K. E., Rice, J. E., Sanchez, J. A., and Wangh, L. J. (2002) QuantiLyse: reliable DNA amplification from single cells. *Biotechniques* **32(5),** 1106–1111.
23. Lewis, C. M., Pinel, T., Whittaker, J. C., et al. (2001) Controlling misdiagnosis errors in preimplantation genetic diagnosis: a comprehensive model encompassing extrinsic and intrinsic sources of error. *Hum. Reprod.* **16,** 43–50.

# 9

## Comparative Genomic Hybridization on Single Cells

### Lucille Voullaire and Leeanda Wilton

### Summary

Comparative genomic hybridization (CGH) is a molecular cytogenetic technique developed for the analysis of chromosome imbalance in tumors and constitutional chromosome abnormalities. It is based on the analysis of genomic DNA and has the advantage over conventional karyotyping in that it does not require that metaphase chromosomes be obtained from the test material. The application of CGH to single cells requires whole-genome amplification of the DNA to provide sufficient DNA for use as a test sample. This approach has been used successfully to identify aneuploidy in single fibroblasts, amniocytes, and buccal cells that were known to be trisomic. CGH can also identify chromosome errors in single blastomeres from early embryos and in first polar bodies. We have analyzed biopsied blastomeres from embryos conceived by in vitro fertilization using CGH in a clinical preimplantation diagnostic program in which euploid embryos are selected for transfer. This has resulted in established pregnancies in patients with recurrent implantation failure.

**Key Words:** Preimplantation diagnosis; comparative genomic hybridization; aneuploidy; blastomere; DOP-PCR; universal DNA amplification.

### 1. Introduction

Comparative genomic hybridization (CGH) is used to examine the entire genome for changes in DNA sequence copy number owing to gain (e.g., trisomy) or loss (e.g., monosomy) *(1–4)*. With CGH two differentially labeled genomic DNA samples are simultaneously hybridized to normal metaphase chromosomes in the presence of Cot-1 DNA, which is used to block the repetitive sequences. The test sample is usually labeled with a green fluorochrome, and the reference (normal DNA) sample is labeled with a red fluorochrome. The basic assumption in CGH is that the hybridization kinetics of the test and reference DNA are independent, so the ratio of binding of the DNA is proportional to the ratio of the copy numbers of the sequences in the DNA samples at a specific locus. The relative fluorescent intensities of the test DNA to the reference

From: *Methods in Molecular Medicine: Single Cell Diagnostics: Methods and Protocols*
Edited by: A. Thornhill © Humana Press Inc., Totowa, NJ

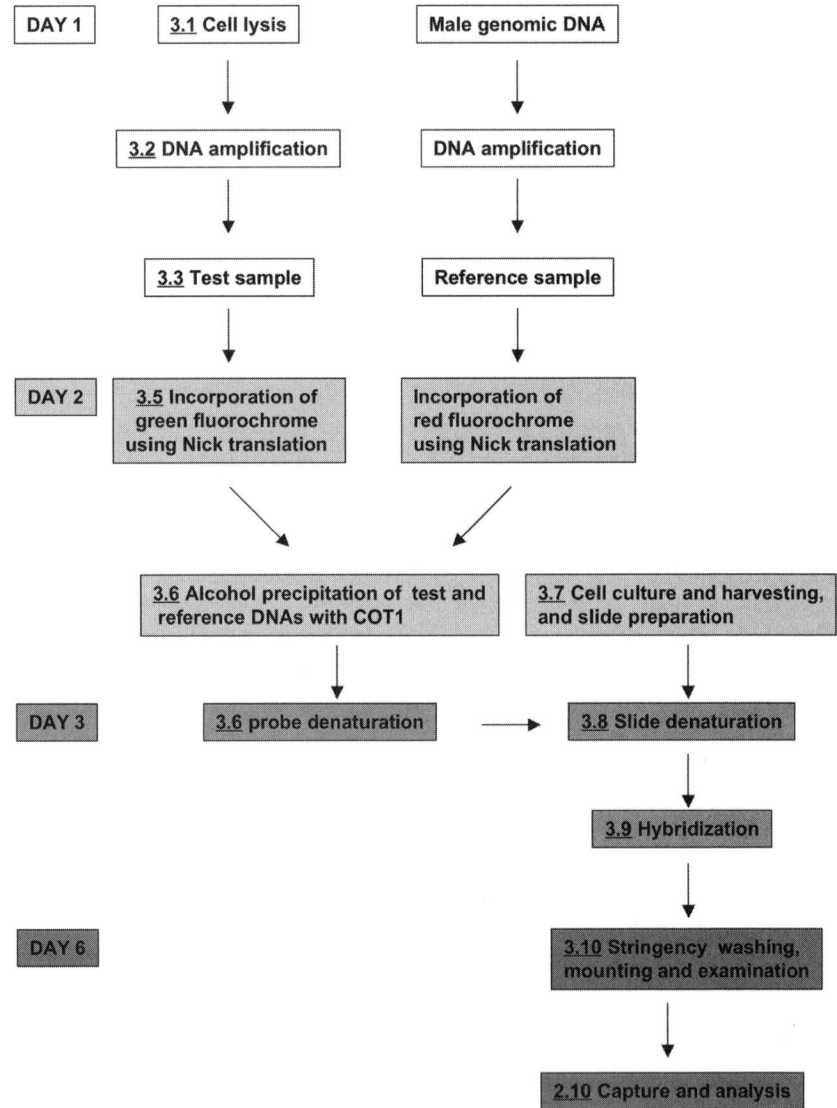

Fig. 1. Flow diagram of SC-CGH method illustrating timescale of entire process from cell lysis to image capture and analysis.

DNA hybridized to the metaphase chromosomes are used to determine the chromosome regions with changed copy number. In the method for single cell CGH (SC CGH) (*see* **Fig. 1**) described herein, the total DNA from a single cell is amplified in a polymerase chain reaction (PCR) using a degenerate oligonucleotide primer (degenerate oligonucleotide–primed [DOP]-PCR) (*see* **Notes 1** and **3–7**). The PCR product is labeled and used as the test sample. The reference DNA (normal male genomic DNA) is also amplified, the PCR product is differentially labeled, and this is used as a reference sample.

The success of this procedure is dependent on efficient amplification of the single cell DNA and effective hybridization of the DNA to metaphase template chromosomes. Care must be taken to avoid contamination when using a universal primer and low template amounts of DNA in a PCR. Preparation for the PCR should be carried out in a designated work area that can be ultraviolet (UV) irradiated. The reaction solutions must be tested and demonstrated to be free of contamination prior to each experimental run. Negative (water) and positive (genomic DNA) controls should be included in each PCR experiment.

Although the resolution of CGH using extracted genomic DNA can be as sensitive as 2 Mb *(2)*, a wider deviation of the profile is observed with PCR-amplified DNA and particularly with DNA amplified from a single cell. The resolution of SC-CGH has been estimated to be about 40 Mb *(3)*. We have found that SC-CGH is able to detect monosomy and trisomy for all chromosomes as well as unbalanced translocations involving segment loss or gain of >40 Mb of DNA.

## 2. Materials

### 2.1. Handling and Lysis of Single Cells

1. Finely pulled glass pipet with a flame-polished tip for handling the single cells.
2. Human Tubal Fluid (Irvine) with 10% human serum added (*see* **Note 2**).
3. Tissue culture oil (In Vitro Fertilization, Turnbull, CT).
4. Lysis buffer: 200 m*M* KOH (analytical grade; BDH, www.bdh.com).
5. Neutralization buffer: 500 m*M* Tris-HCl, pH 8.3; 300 m*M* KCl; 200 m*M* HCl (all Analar grade; BDH) (*see* **Note 3**).

### 2.2. Amplification of DNA Using DOP-PCR

1. 2 m*M* dNTP solution (Roche, www.roche-applied-science.com).
2. 25 m*M* MgCl$_2$ (Perkin Elmer,http://las.perkinelmer.com).
3. 750 µ*M* Primer 6MW *(11)* [5'-CCG ACT CGA GNN NNN NAT GTG G-3'] (Sigma-Genosys, www.sigmaaldrich.com) (*see* **Note 4**).
4. Taq polymerase (5 U/µL) (Perkin Elmer).
5. PCR buffer without MgCl$_2$ (10X) (Perkin Elmer) for amplification of reference DNA.
6. Normal male reference DNA (1 ng/µL) (prepared in-house).

### 2.3. Gel Electrophoresis to Detect PCR Product

1. Small electrophoresis gel equipment.
2. 11-Well comb.
3. Agarose MP (Roche).

4. Ethidium bromide (10 mg/mL) (Sigma-Aldrich).
5. TBE buffer: 89 m$M$ Tris; 89 m$M$ boric acid; 2 m$M$ EDTA, pH 8.0 (all analytical grade; BDH).
6. DNA molecular marker (50 ng/µL), e.g., marker X (Roche) diluted in loading buffer.
7. Nescofilm (Bando, Kobe, Japan).
8. 6X Gel loading buffer: 0.25 g of bromophenol Blue (Sigma-Aldrich), 30 mL of glycerol, 70 mL of 5X TBE.

### 2.4. Alcohol Precipitation of DNA

1. 3 $M$ Sodium acetate (Analar grade; BDH) stored at room temperature.
2. 100 and 70% ethanol stored at –20°C.
3. Nuclease-free water (Vysis, www.vysis.com).

### 2.5. Incorporation of Fluorochrome Using Nick Translation

1. Nick translation kit (Vysis).
2. Red and green fluorochromes (Spectrum Red-dUTP and Spectrum Green-dUTP; Vysis).

### 2.6. Preparation and Denaturation of Probe

1. Cot–1 DNA (Roche).
2. 3 $M$ Sodium acetate.
3. 100% Ethanol at –20°C.
4. Hybridization buffer: 50% deionized formamide (formamide [Sigma-Aldrich] is deionized to pH 7.0 over AG 501-X8 Resin [Bio-Rad]), 2X SSC, 10% dextran sulfate (stock solution of 50% dextran sulfate made by dissolving in water at 60°C; store at 4°C), (8.82 gm/L Na$_3$ citrate; 17.53 g/L NaCl) final solution is pH 7.0.

### 2.7. Preparation of Metaphase Slides

1. Sterile venous whole blood sample (5 mL) from normal male.
2. Sterile 25-cm$^2$ Falcon flasks for culturing (www.bdbiosciences.com).
3. Sterile culture medium containing RPMI, 10% fetal calf serum, glutamine, and antibiotics.
4. Phytohemagglutinin (PHA) (Gibco; Invitrogen, www.invitrogen.com).
5. Centrifuge tubes (Falcon).
6. 0.1 $M$ Thymidine (Sigma-Aldrich).
7. Colcemid (10 µg/mL) (Gibco, Invitrogen).
8. 0.56% KCl (Sigma-Aldrich).
9. Fixative: 1 part acetic acid (analytical grade; BDH):3 parts methanol (analytical grade C Biolab, www.biolabgroup.com).
10. Gilson Pipetman (20 µL) (www.gilson.com) and filtered tips (Axygen, www.axygen.com).
11. Slides (Superfrost® Plus) (Menzel-Glaser, www.menzel.de).
12. Sealable slide storage containers.

13. Dessicant.
14. Phosphate-buffered saline (PBS) (pH 7.4) solution made from buffer tablets (Oxoid, www.oxoid.com).
15. Ethanol series (70, 85, and 100%).

### 2.8. Denaturation of Slides

1. Heat-resistant Coplin jar.
2. Formamide denaturing solution: 70% deionized formamide and 30% 2X SSC, pH 7.0.
3. Water bath at 73°C.
4. Coplin jar containing 70% ethanol at –20°C.
5. Coplin jars containing 70, 85, and 100% ethanol at room temperature.

### 2.9. Hybridization, Stringency Washing, and Mounting of Slides

1. $CO_2$-gassed incubator set at 37°C.
2. Humidified container such as a sealable plastic box with damp filter paper in the base.
3. 1X SSC.
4. 2X SSC.
5. Ethanol (70, 85, and 95%).
6. Vectashield (Vector, www.vectorlabs.com) containing 0.5 µg/mL of 4,6-diamidino-2-phenylindole (DAPI) (Sigma-Aldrich).

### 2.10. Image Capture, CGH Analysis, and Interpretation

1. Epifluorescence microscope with digital camera attached to computer imaging system. Filters appropriate to the excitation and emission qualities of each fluorochrome are required (DAPI: 350-nm excitation/460-nm emission; Spectrum Green: 480 nm/535 nm; Spectrum Red: 560 nm/630 nm) (Chroma, www.chroma.com).

## 3. Methods

### 3.1. Handling and Lysis of Single Cells

Individual cells are washed and then transferred into an alkaline lysis buffer in a PCR tube. Tissue culture mineral oil is added to prevent evaporation and the sample is heated to ensure complete lysis. The lysis solution is then neutralized (*see* **Note 5**).

1. Add 5 µL of lysis buffer to the required number of 0.2-mL clean, sterile PCR tubes.
2. Wash individual blastomeres in small drops of clean culture medium containing human serum albumin.
3. Transfer a single blastomere into each PCR tube in a minimum volume (<0.5 µL) of culture medium.
4. Add 10 µL of tissue culture oil above the buffer.
5. Heat the tube to 65°C for 10 min.
6. Cool quickly on ice, add 5 µL of neutralization buffer, and spin briefly. Cells may be stored at this stage at –20°C for up to 3 mo.

### 3.2. Amplification of DNA Using DOP-PCR

Amplification of the entire genome by PCR can be achieved using a degenerate universal primer *(8)*. The recommended primer has a random hexamer sequence flanked by a 6-bp defined sequence at the 3' end and a 10-bp cloning sequence at the 5' end. The primer is designed to hybridize every 3000 bp on average, and amplification results in fragments varying in size from 300 to 2000 bp.

1. Make up the required amount of PCR reaction mix based on the number of tubes required for the test and for reference samples and controls allowing one extra quantity of PCR mix for each 1–10 tubes. Combine the following for the PCR reaction mix for reference samples: 32.5 µL of water, 1 µL (1 ng/µL) of genomic DNA, 5 µL of buffer, 5 µL of dNTP, 5 µL of $MgCl_2$, 1 µL of primer, and 0.5 µL of Taq, for a total volume of 50 µL.
2. Dispense 50 µL into individual tubes.
3. For the single cell test samples, replace the PCR buffer and 5 µL of water with the lysis and neutralization buffers. Combine the following for the PCR reaction mix for single cell test samples: 28.5 µL of water, 5 µL of dNTP, 5 µL of $MgCl_2$, 1 µL of primer, and 0.5 µL of Taq, for a total volume of 40 µL.
4. To each test sample tube containing lysis and neutralization buffers, add 40 µL of PCR reaction mix and spin briefly.
5. Carry out PCR reaction using the following program:
   a. Cycle 1: 94°C for 5 min (DNA denaturation).
   b. Cycle 2: 94°C for 30 s (DNA denaturation); 30°C for 1 min (primer annealing); RAMP 25% (i.e., 1°C/4 s) between 30 and 72°C; 72°C for 1 min (extension) for 8 cycles.
   c. Cycle 10: 94°C for 30 s; 56°C for 1 min; 72°C for 1 min with a 10-s extension/cycle for 35 cycles.
   d. Cycle 46: 72°C for 7 min.
   e. Cycle 47: Hold at 4°C.

The PCR product can be stored at –20°C.

### 3.3. Gel Electrophoresis to Detect PCR Product

A sample of each PCR product is run on an electrophoresis gel containing ethidium bromide. This enables an assessment of the quality of the PCR product and its suitability for CGH as well as ensures that control samples have given appropriate results.

1. Make a 1% agarose gel in a small gel tray (700 mg of agarose in 70 mL of TBE buffer, 1 µL of ethidium bromide) using an 11-well comb. Remove the tubes from the PCR machine and spin briefly.
2. In the first well add 2 µL of marker DNA.
3. Place at intervals 1-µL samples of loading buffer onto Nesco film. For each tube remove 3 µL of PCR product from the tube and add to the 1 µL of loading buffer. Mix and place the 4 µL into the next well.

4. Repeat for all samples noting the order of the tubes and wells.
5. Run gel at 100 V, 50 mA for 45 min. Examine under UV light (*see* **Note 6**).

### 3.4. Alcohol Precipitation of DNA

The PCR product is prepared for labeling by ethanol precipitation in the presence of a high salt concentration. This allows removal of the PCR reagents and concentration of the DNA sample.

1. Before starting, label clean, sterile 1.5-mL Eppendorf tubes corresponding to the test and reference samples.
2. Set up PCR tubes and Eppendorf tubes in corresponding order. Be careful to check correspondence at transfer.
3. Spin the PCR reaction tube to collect the product. Carefully remove the PCR product (47 µL) from below the oil layer. Transfer to the correspondingly labeled Eppendorf tube.
4. Add 5 µL of 3 *M* NaAc (i.e., the equivalent of 10% of the original volume) and 125 µL of 100% ethanol (i.e., 2.5X volume).
5. Leave the tubes at –20°C for 2–16 h.
6. Place the tubes with the hinge toward the outer rim of a refrigerated centrifuge so that the product is spun against the hinged side of the tube. Spin the tubes at 16,000*g* for 30 min at 4°C.
7. Carefully draw up the supernatant by sliding the pipet tip to the side of the tube opposite the hinge, and discard.
8. Replace with 70% cold ethanol (–20°C). Do not disturb the pellet.
9. Spin at 14,000 rpm for 5 min. Remove the supernatant. Spin briefly to collect the remaining ethanol and then carefully remove as much of this as possible.
10. Dry the pellet by leaving the tube open at room temperature for 30 min, or at 37°C for 15 min (*see* **Note 7**). The DNA can be stored at –20°C after redissolving in 10 µL of nuclease-free water. This volume of water will need to be accounted for in the PCR reaction mix (*see* **Subheading 3.5., step 2**).

### 3.5. Incorporation of Fluorochrome by Nick Translation

The PCR product is labeled by incorporation of a fluorochrome-labeled dUTP in a nick translation reaction. The test and reference DNA samples are differentially labeled with red and green fluorochromes.

1. A nick translation kit is used and reagents need first to be made up according to the manufacturer's instructions. The reagents are stored at –20°C.
2. Hold each tube of dried PCR sample on ice and add the reagents from a PCR master mix containing all reagents except the DNA. Make up the PCR master mix according to the number of tubes. Combine the following for the PCR reaction mix per tube: 17.5 µL of water (allow for the volume of the DNA sample if this is already in solution), 2.5 µL of 0.2 m*M* Spectrum green dUTP for test DNA or 0.2 m*M* Spectrum Red dUTP for reference DNA, 5.0 µL of 0.1 m*M* dTTP, 10.0 µL of

0.1 m*M* dNTP, 5.0 µL of 10X nick translation buffer, and 10.0 µL of nick translation enzyme, for a total volume of 50.0 µL.
3. Vortex the tube briefly.
4. Incubate at 15°C in a water bath for 6–8 h.
5. Stop the reaction by heating in a 70°C water bath for 10 min.
6. Place on ice. The sample can be stored at –20°C.
7. Check size of the DNA fragments (<1000 bp with the majority ~500 bp) by removing 3 µL of sample (mix with 1 µL of loading buffer) and run on a 1% electrophoresis gel as described under **Subheading 3.3** (*see* **Note 8**).

## 3.6. Preparation and Denaturation of Probe

The labeled test and reference DNA together with human Cot-1 DNA are
alcohol precipitated, dried, and resuspended in hybridization buffer. The probe
is denatured at 75°C for 8 min.

1. To each reaction tube (containing the labeled DOP-PCR product from a single cell)
   add 30 µL of Cot–1, 30 µL of labeled DOP-PCR-amplified reference DNA, 11 µL
   of 3 *M* NaAc, and 300 µL of 100% ethanol. Alcohol precipitate as described under
   **Subheading 3.4**.
2. Resuspend the dry pellet in 10 µL of hybridization buffer. Vortex briefly, spin, and
   allow to dissolve at 37EC for 30 min. The probe sample can be stored in the dark
   at –20°C for up to 12 mo.
3. Denature the probe by placing in a water bath at 73EC for 10 min. Place on ice and
   then spin to collect the contents.

## 3.7. Preparation of Metaphase Template Slides

The quality of the metaphase template slides is crucial to the success of the
CGH (**a**). Slides with a high number of well-spread metaphases of a consistent length corresponding to approx 400 bands are required. Slides suitable
for CGH can be obtained from Vysis (www.vysis.com). Alternatively, slides
can be prepared in the laboratory from cultured peripheral human blood
obtained from a normal male donor after standard culturing and harvesting
procedures. Thymidine synchronization of cell division to increase the
mitotic index and the length of the chromosomes is recommended. Other synchronization methods (*10*) may result in metaphases with excessively long
and twisted chromosomes. Slides are prepared by dropping 12 µL of cell suspension onto a flat slide at two spots on the slide. After drying, the slides are
passed through an ethanol series to remove the acid fixative. This helps con-
serve the integrity of the chromosomal DNA by minimizing depurination of
the DNA on the slide (*11*).

It is preferable for consistency to prepare a large number of slides that can
be stored until required at –20°C in a slide container in a sealed bag with desiccant. Slides are left to age at room temperature overnight before either imme-

diate use or storing. Each batch of slides will need to be assessed for the quality of hybridization and to determine a suitable denaturation time.

### 3.7.1. Cell Culture

1. Add 200 µL of fresh blood to 8 mL of complete medium containing PHA in a 25-cm$^2$ Falcon plastic flask.
2. Incubate the flask upright at 37°C for 48 h.
3. Add 80 µL/100 m$M$ thymidine to each container and return to the incubator.
4. After 16 h transfer each sample to a separate sterile centrifuge tube.
5. Spin at 1300 rpm for 10 min.
6. Add 8 mL of fresh RPMI.
7. Mix and spin at 2400$g$ for 10 min.
8. Remove the supernatant and add back 8 mL of complete medium.
9. Incubate for a further 6 h at 37°C.

### 3.7.2. Harvesting of Cells

1. Add 100 µL of colcemid to each culture and return to the incubator for 20 min.
2. Spin the tube and remove the supernatant.
3. Resuspend the cells using a vortex. Add 8 mL of warmed 0.56% KCl and mix.
4. Incubate at 37°C for 10 min.
5. Spin at 2400$g$ for 8 min.
6. Remove the supernatant leaving ~0.5 mL of KCl.
7. Resuspend using a vortex to give a single cell suspension. While maintaining the tube on the vortex, add 1.0 mL of fixative dropwise to the cells.
8. Add a further 7 mL of fixative. Replace the cap and rotate to ensure an even single cell suspension.
9. Spin at 2400$g$ for 8 min. Remove the supernatant. Add 8 mL of fresh fixative.
10. Repeat **step 9**.
11. Spin at 2400$g$ for 8 min. Remove the supernatant. Add ~1 mL of fixative so as to achieve a slightly cloudy cell suspension. Proceed to **Subheading 3.7.3.** for preparation of slides. Alternatively, cell suspension can be stored in 5–10 mL of fixative at –20°C for up to 6 mo in a tightly capped sealed container.

### 3.7.3. Preparation of Slides

1. Make a slide by dropping 12 µL of cell suspension onto a flat slide at two separate positions. Chromosome spreading may be enhanced by dropping a further 12 µL of fresh fixative onto the dropped cells just prior to the cells drying (*see* **Note 9**).
2. Leave the slides to air-dry and examine the quality of the chromosome spreading using phase contrast microscopy. There should be a high mitotic index with 5–10 metaphase cells per high-power field. A high-quality preparation is one in which the chromosomes are straight and separated with minimum overlaps.
3. Wash the slides for 2 min at room temperature in PBS and dehydrate in an ethanol series (70, 85, and 100%).

4. Leave the slides to age overnight at room temperature.
5. Store the slides in sealed slide containers containing desiccant at –20°C.

## 3.8. Denaturation of Slides

Chromosomal DNA is denatured *in situ* by placing slides in a 70% formamide solution at 73°C.

1. Remove the slides in the storage box from the freezer and allow to come to room temperature before removing the slides from the container. Check the slides for metaphase number and quality and then label the slides.
2. Place a Coplin jar containing 45 mL of formamide denaturing solution in a 73°C water bath.
3. When the temperature *in* the Coplin jar has reached 73°C add two slides. Denature for 5 min.
4. Transfer the slides into 70% ethanol at –20°C.
5. Dehydrate through an ethanol series at room temperature (70, 85, and 95% ethanol) for 1 min each.
6. Place on a heat tray at 42°C to dry.
7. Do not denature more than two slides at a time, and allow the temperature of the denaturation solution to return to 73EC before adding fresh slides.

## 3.9. Hybridization, Stringency Washing, and Mounting of Slides

Incubation at 37°C for 3 d is required for maximum hybridization of the probe DNA to the template chromosomal DNA. Incubation for less than 3 d results in weaker hybridization. The slides are subsequently washed to remove excess probe and then mounted. A fluorescent DNA counterstain is included in the mountant.

1. Add 4 μL of probe to the denatured slide. Cover with a 13-mm round cover slip and seal with rubber cement. Allow the cement to set before transferring the slide to a prewarmed moist, sealable chamber. Incubate at 37°C for 3 d.
2. To stringency wash the slides, place 50 mL of 1X SSC in each of two Coplin jars and warm to 70°C in a water bath.
3. Remove the rubber cement from around each cover slip, immerse the slide in 2X SSC, and carefully slide off the cover slip.

---

Fig. 2. CGH profile generated following cohybridization of amplified DNA from single cell biopsied from early human embryo with amplified normal male DNA. The cell shows multiple abnormalities including both chromosome loss and gain. The profile ratios of 0.5 to 1.5 are indicated by straight lines in increments of 0.25. The majority of the chromosomes show a normal ratio of 1. Chromosomes 10 and 17 show marked deviation to the right (>1.25), indicating that there is additional chromosome 10 and 17 DNA in the test sample, whereas chromosomes 16 and 18 show deviation to the left (<0.75), indicating relative deficiency of test DNA for these chromosomes.

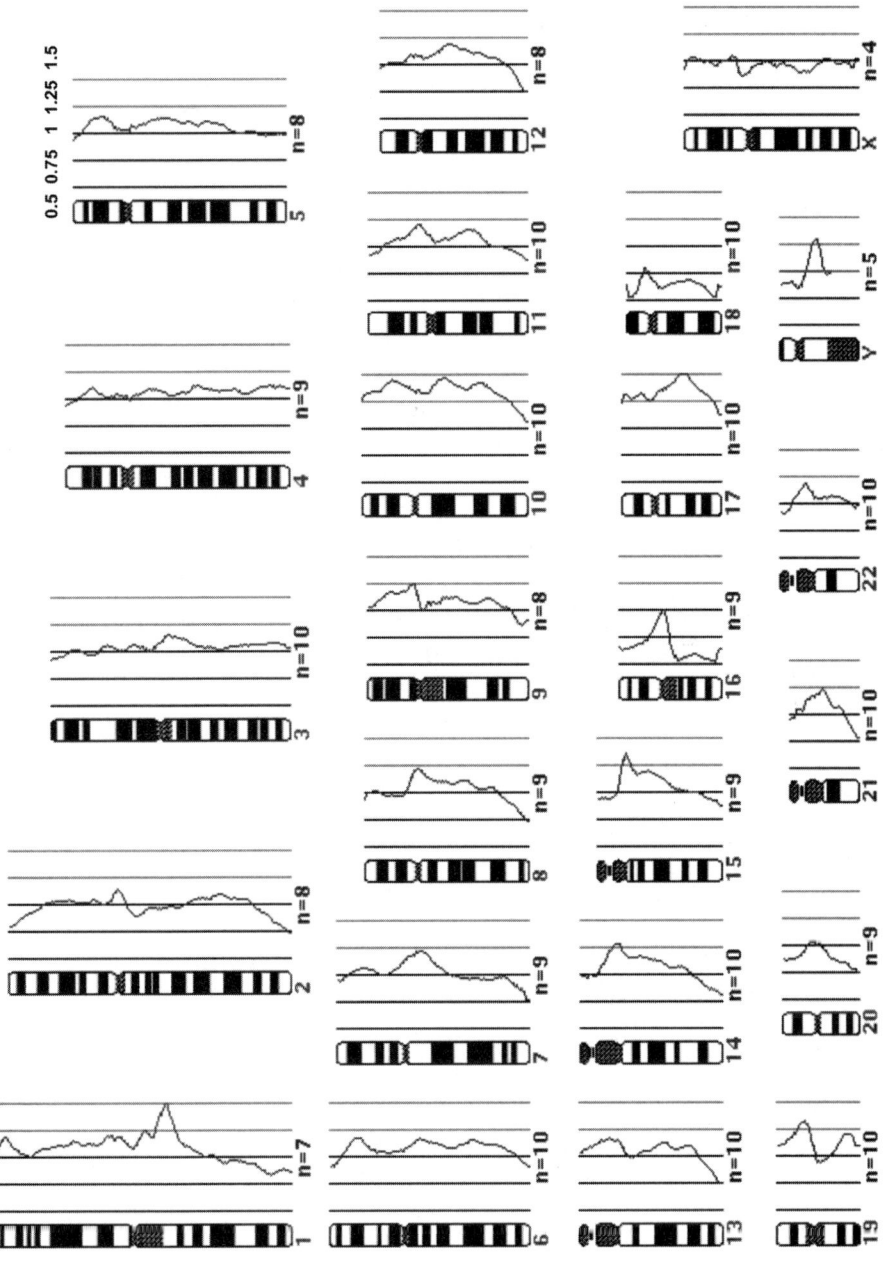

4. Wash the slides twice in 1X SSC at 70°C for 5 min each.
5. Transfer the slides to 2X SSC at room temperature, and then dehydrate through an alcohol series (70, 85, and 95% ethanol) at room temperature. Air-dry in the dark.
6. Mount the slides with Vectashield containing 0.4 μg/μL of DAPI under a cover slip.

### 3.10. Capture, Analysis, and Interpretation

The slides are examined using an epifluorescence microscope with filters that are appropriate to the fluorochromes used. Digital capture of fluorescence and CGH analysis are carried out using computer software. We have used Cytovision (Applied Imaging), but similar software is available from other manufacturers such as Metasystems (Carl Zeiss). Essentially gray-field images of blue (chromosome counterstain), green (test DNA), and red (reference DNA) fluorescence are digitally captured. The images are pseudocolored for display. The test and reference DNA are normalized to give a ratio of 1 across the entire metaphase. The relative hybridization along each chromosome is determined and presented as a profile against an ideogram of the chromosome. Loss of DNA in the test sample will result in deviation of the profile to the left into the red, and excess test DNA sequences will result in deviation of the profile to the right into the green. With Cytovision software, deviation of the profile to <0.75 or >1.25 is generally regarded as indicating significant variation of the test sample.

Metaphase preparations are examined using an epifluorescence microscope with appropriate filters. Image capture and CGH analysis is carried out using computer software such as the Cytovision CGH program (Applied Imaging). Gray tone images are captured of the stained metaphase and of the red and green fluorescent hybridization. The CGH software determines the average red-to-green fluorescent ratio for each chromosome. In regions where the DNA sequence copy number of the test and the reference DNA are identical, the normalized fluorescent ratio is expected to be 1.0; where the relative DNA sequence copy number is greater in the test, the ratio is >1.0; and where the relative DNA sequence copy number is less in the test, the ratio is <1.0. Using Cytovision software deviations of the ratio below or above the global thresholds of 0.75–1.25 is regarded as representing significant under- or overrepresentation of the DNA sequence copy number in the test sample (**Fig. 2**). Major deviations of the profile can occur at the telomeres owing to weak DNA hybridization in these regions and around the centromeres owing to the low levels and variable amounts of repetitive DNA still present despite inclusion of Cot–1 in the hybridization probe mix (*see* also **Note 10**).

## 4. Notes

1. Other single cell amplification methods might be considered. A recent development has been the use of phi polymerase *(12)*, but in our hands this method does not

appear to give a sufficiently even amplification of the DNA from a single cell to be suitable for use with SC-CGH nor is it recommended for this purpose by the manufacturer (Amersham Biosciences, www.amershambiosciences.com). Linker ligation followed by DNA amplification was recently used for low-template DNA and successfully applied to SC-CGH *(13)*.

2. Single blastomeres can be difficult to handle, because they easily stick to glass and plastic surfaces. This problem can be minimized by using a protein source, such as 10 mg/mL of human albumin or 10% human serum, in the medium.

3. The lysis and neutralization buffers together are equivalent to the standard PCR buffer for Taq polymerase (Perkin-Elmer).

4. The degenerate (high-performance liquid chromatography–purified) primer used in DOP-PCR varies among different manufacturers and even among different batches from the same manufacturer. Consequently, each batch of primer must be assessed on the basis of the intensity and dynamic quality of the CGH. Owing to batch variation, the final primer concentration is determined empirically. The use of excess primer can lead to inhibition of the reaction and insufficient primer results in reduced PCR product observed with the positive control. The primer concentration used is determined by these experimental criteria.

5. Other lysis solutions such as digestion with SDS proteinase K have been successfully used by other groups *(7)*.

6. On the gel the negative control should show a completely clear lane. The positive genomic control and test DNA from a single cell should show a smear with a range of approx 200–2000 bp. Reference DNA and single cells from sources other than blastomeres frequently show a band at 450 bp within the smear. Blastomeres show a smear but also show several additional bands, particularly at 600, 1250, and 1650 bp, corresponding to mitochondrial DNA *(6)*. The presence of just these bands and no smear is evidence of an anucleate cell. The presence of random bands is indicative of contamination. This PCR product obtained from random contamination in a negative control does not generally show hybridization in a control CGH experiment, so it is unlikely to affect the final CGH analysis. However, a repeated pattern suggests that the solutions are contaminated and need to be replaced. A long smear in the negative control, i.e., starting in the well and extending to a few hundred base pairs, does not reflect contamination but is most likely the result of formation of concatamers of the primer. This can occur inconsistently with the same set of reagents in a series of PCR reactions. Reducing the concentration of the primer may reduce the formation of this smear in the negative control, but it might also reduce the product in the positive control (*see* **Note 4**).

7. It is important that the DNA product be dry before proceeding with the next step. However, care must also be taken not to dry the precipitate excessively, because it can be difficult to redissolve.

8. Probe size is important for a successful CGH experiment *(9)*. The probe size will be reduced if the temperature is allowed to increase above 15°C or if the batch of enzyme is particularly active. The probe size can be checked at 6 h. Place the tubes on ice to stop the reaction while running the gel. If the DNA is still in large fragments,

place the tubes back at 15°C for another 2 h or longer. The concentration of enzyme or time of incubation can be reduced if the enzyme batch appears to be particularly active. If the DNA has been stored in solution, adjust the amount of nuclease-free water to keep the total reaction volume at 50 µL. The fluorochrome-labeled dUTPs and labeled DNA must be protected from the light.

9. Cell spreading is influenced by many factors *(9)*. A propensity for cells to break too quickly on spreading resulting in loss of chromosomes can be improved by leaving newly fixed cells at –20°C overnight. The humidity of the working area, the ratio of the reagents in the fixative, dropping onto a wet slide, and altering the temperature of the slide will all affect the spreading of the chromosomes. The DAPI banding following a denaturation step (*see* **Subheadings 3.8.** and **3.9.**) can be used to assess slide suitability. The metaphase chromosomes should show clear bands following denaturation when stained with DAPI. Bright centromeric banding and poor "G banding" is indicative of overdenaturation and weak centromeric banding is indicative of underbanding *(9)*.

10. The use of SC-CGH in the investigation of single blastomeres from human embryos has shown that there can be a wide variety of chromosomal abnormalities including trisomy or monosomy involving a single chromosome, but also abnormalities involving imbalance of many chromosomes *(5–7)*. Within the limits of resolution of the technique, partial chromosome imbalance following chromosome breakage can also be detected using SC-CGH and has been repeatedly observed in individual blastomeres *(5–7)*. However, the artefactual deviation of the profile at the telomeres precludes SC-CGH for the analysis of imbalance in blastomeres arising from many balanced parental translocations *(14)*. Complex abnormality that frequently occurs in blastomeres can be observed following SC-CGH but often cannot be interpreted in terms of specific chromosomal involvement, because the profile is distorted by the extremely uneven involvement of individual chromosomes *(6)*.

## REFERENCES

1. Kallionemi, O.-P., Kallionemi, A., Sudar, D., et al. (1993) Comparative genomic hybridisation: a rapid new method for detecting and mapping DNA amplification in tumours. *Cancer Biol.* **4,** 41–46.

2. Ness, G.O., Lybaek, H., and Houge, G. (2002) Usefulness of high-resolution comparative genomic hybridization (CGH) for detecting and characterizing constitutional chromosome abnormalities. *Am. J. Med. Genet.* **113,** 125–136.

3. Voullaire, L., Wilton, L., Slater, H., and Williamson, R. (1999) Detection of aneuploidy in single cells using comparative genomic hybridisation. *Prenat. Diagn.* **19,** 846–851.

4. Wells, D., Sherlock, J.K., Handyside, A.H., and Delhanty, J. D. A. (1999) Detailed chromosomal and molecular genetic analysis of single cells by whole genome amplification and comparative genomic hybridisation. *Nucleic Acids Res.* **27,** 1214–1218.

5. Voullaire, L., Slater, H., Williamson, R., and Wilton, L. (2000) Chromosome analysis of blastomeres from human embryos by using comparative genomic hybridisation. *Hum. Genet.* **106,** 210–217.

6. Voullaire, L., Wilton, L., McBain, J., et al. (2002) Chromosome abnormalities identified by comparative genomic hybridization in embryos from women with repeated implantation failure. *Mol. Hum. Reprod.* **8,** 1035–1041.

7. Wells, D. and Delhanty, J. D. A. (2000) Comprehensive chromosomal analysis of human preimplantation embryos using whole genome amplification and single cell comparative genomic hybridization. *Mol. Hum. Reprod.* **6,** 1055–1062.

8. Telenius, H., Pelmear, A. H. P., Tunnacliffe, A., et al. (1992) Cytogenetic analysis by chromosome painting using DOP-PCR amplification of flow sorted chromosomes. Genes Chromosomes Cancer **4,** 257–263.

9. Watt, J. L. and Stephen, G. S. (1986) Lymphocyte culture for chromosome analysis, in *Human Cytogenetics—A Practical Approach* (Rooney, D. E. and Czepulkowski, S. H., eds.), IRL Press, Oxford, UK, pp. 39–56.

10. Mezzanotte, R., Vanni, R., Flore, O., Ferucci, L., and Sumner, A. T. (1988) Ageing of fixed cytological preparations produces degradation of chromosomal DNA. *Cytogenet. Cell Genet.* **48,** 60–62.

11. Kalliomieni, O.-P., Kallioniemi, A., Piper, J., et al. (1994) Optimizing comparative genomic hybridization for analysis of DNA sequence copy number changes in solid tumours. *Genes Chromosomes Cancer* **10,** 231–243.

12. Klein, C. A., Schmidt-Kittler, O., Schardt, J. A., Pantel, K., Speicher, M. R., and Riethmüller, G. (1999) Comparative genomic hybridization, loss of heterozygosity, and DNA sequence analysis of single cells. *Proc. Natl. Acad. Sci. USA* **96,** 4494–4499.

13. Lage, J. M., Leamon, J. H., Pejovic, T., et al. (2003) Whole genome analysis of genetic alterations in small DNA samples using hyperbranched strand displacement amplification and array-CGH. *Genome Res.* **13,** 294–307.

14. Malmgren, H., Sahlen, S., Inzunza, J., et al. (2002) Single cell CGH analysis reveals a high degree of mosaicism in human embryos from patients with balanced structural chromosome aberrations. *Mol. Hum. Reprod.* **8,** 502–510.

# 10

## Generation of Amplified RNAs and cDNA Libraries from Single Mammalian Cells

### James Adjaye

### Summary

With the near completion of the human genome sequencing effort, it is now possible to analyze the expression of the entire human gene complement. However, a major obstacle in performing such analysis is the ability to successfully generate enough cDNA or amplified RNA
from a limited number of cells, such as biopsies, blood smears, cells obtained by laser capture
microscopy, and preimplantation embryonic cells and germ cells. Because these samples yield
extremely small amounts of RNA, reproducible methods are needed to amplify this RNA while
maintaining the original message profile. A detailed description is given for generating pools of
cDNA libraries containing a high proportion of cDNAs enriched with 5'-coding sequences from
as little as 1 ng of total RNA using a modified switching mechanism at 5' end of RNA transcript
protocol. In addition, the T7-promoter-linked double-stranded cDNAs can be in vitro transcribed
linearly using T7-RNA polymerase to generate amplified RNA that is mRNA derived. The cDNA
pools can be used directly for gene-specific reverse transcriptase polymerase chain reaction or
processed for ligation into vectors of choice whereas the amplified RNA can be used for microarray-based expression profiling.

**Key Words:** Preimplantation embryos; mRNA isolation; reverse transcriptase polymerase
chain reaction; in vitro transcription; expression profiling; cDNA libraries.

## 1. Introduction

The ultimate aim in clinical diagnosis, disease-related research, and developmental biology is to be able to carry out analysis with as few cells as possible, preferably single cells. However, a major obstacle to such analysis is the
minute amount of RNA obtainable from such samples. The total RNA content
of a single mammalian cell is in the range of 20–40 pg *(1)* and only 0.5–1.0 pg
of this is mRNA. Consequently, any analysis of gene expression at the level of
single cells must be capable of dealing with a total of only $10^5$–$10^6$ mRNA molecules; therefore, amplification is unavoidable. The first step toward this goal is

From: *Methods in Molecular Medicine: Single Cell Diagnostics: Methods and Protocols*
Edited by: A. Thornhill © Humana Press Inc., Totowa, NJ

mRNA isolation. This is followed by double-stranded cDNA synthesis employing whole-genome reverse transcriptase polymerase chain reaction (RT-PCR). An additional option is to use generated double-stranded cDNA incorporating the T7-promoter sequence as template for in vitro transcription to generate million-fold amplifications of the original mRNA population *(2)*. As an illustration, a detailed description of an improved and reproducible protocol for isolating mRNAs from as little as 1, 5, 10, and 50 ng of total RNA is given. These concentrations are equivalent to 0.05, 0.25, 0.5, and 2.5 ng of mRNA, assuming that the composition of mRNA within a cell is 5% of total RNA. Double-stranded cDNAs are generated by a modified switching mechanism at 5' end of RNA transcript (SMART) protocol and can be used directly for gene-specific RT-PCR or processed for ligation into vectors of choice *(3–10)*. The amplified RNA, on the other hand, can be used for microarray-based expression profiling *(11 –13)*.

Important features of this method are as follows:

1. All reactions are performed directly on the oligo-(dT) magnetic beads (solid phase), thus minimizing potential losses at each purification step.
2. Use of the SMART oligo tags the 5'-end of the first-strand cDNAs; enables PCR amplification of the cDNAs in their entirety; and eliminates the need for conventional second-strand cDNA synthesis and troublesome time-consuming, adaptor ligation prior to cloning into an appropriate vector.
3. Tagging of the 5' ends of the synthesized cDNAs with the SMART sequence enables the use of rapid amplification of cDNA ends *(14)* to extend partial cDNAs toward their 5' ends.

## 2. Materials

### 2.1. Reagents

1. Superscript II (cat. no. 18064-014; Invitrogen).
2. T4 gene 32 protein (cat. no. 972 983; Roche).
3. RNasin (cat. no. N2511; Promega).
4. IGEPAL CA-630 (cat. no. I-3021; Sigma-Aldrich).
5. Nuclease-free water (cat. no. P1193; Promega).
6. Dynabeads mRNA DIRECT™ Micro kit (cat. no. 610-21; Dynal).
7. Magnetic Particle Concentrator (MPC® ; Dynal).
8. Advantage 2 PCR kit (cat. no. K1910-y; Clontech).
9. MessageAmp™ aRNA kit (cat. no.1750; Ambion).
10. Oligonucleotide primers (*see* **Note 1**):
    a. Primer 1: Oligo(dT)-T7 promoter primer
       5'- <u>TAATACGACTCACTATAGG</u><u>CGGAGG</u>CGG -(dT)$_{20}$ N$_{-1}$N-3'
              **T7 PROMOTER**        *Sfl* **1B**
       (N = A, G, C or T; N$_{-1}$ = A, G, or C)
    b. Primer 2: SMART oligo
       5'-AAGCAGTGGTATCAACGCAGAGTGGCC<u>ATTAT</u>GGCCGGG-3'
                                            *Sfl* **1A**

    c. Primer 3: 5' PCR
    5'-AAGCAGTGGTATCAACGCAGAGT-3'
    d. Primer 4: *b -ACTIN*—Forward
    5'-CGGATGTCCACGTCACACTT-3'
    e. Primer 5: *b -ACTIN*—Reverse
    5'- GTTGCTATCCAGGCTGTGCT-3'
11. pDNR-LIB plasmid vector (cat. no. 6339-1; BD Biosciences Clontech). pTriplEx2 lambda vector (cat. no. PT3003-1; BD Biosciences Clontech).
12. NucleoSpin® Extraction II kit (cat. no. 636972; BD Biosciences Clontech).
13 Single cell lysis buffer (5 m$M$ dithiothreitol [DTT], 10 U of RNasin, 0.8% IGEPAL).
14. Mineral oil (cat. no. M-5904; Sigma-Aldrich).

## 2.2. Equipment

1. Agarose and gel electrophoresis chamber.
2. PCR machine.
3. SpeedVac.

# 3. Methods
## 3.1. Cell Lysis and Isolation of mRNA by Affinity Purification

1. If starting from preimplantation embryos or single cells, place samples in 5 µL of lysis buffer (*see* **Note 2**). (Samples can be snap frozen in liquid nitrogen and stored at –70°C indefinitely.)
2. Add 5 µL of 2X binding buffer to the cell lysate or purified RNA.
3 Mix by gently pipetting, and heat at 65°C for 5 min using a thermal cycler or block.
4. Add 10 µL of oligo-dT magnetic beads prewashed twice with 1X binding buffer (*see* **Note 3**). For each wash, add 50 µL of appropriate buffer and spin briefly prior to placing in magnetic particle concentrator, and then carefully remove the supernatant surrounding the concentrated beads.
5. Mix and then let stand at room temperature for 20–30 min.
6. Wash the beads twice with 50 µL of wash buffer provided in the kit.
7. Wash twice with 1X RT buffer (provided as 5X stock solution RT buffer with the Superscript kit). For each wash, add 50 µL of RT buffer and spin briefly prior to placing in magnetic particle concentrator.
8. Add 2 µL of RNase-free water and proceed with reverse transcription.

## 3.2. First-Strand cDNA Synthesis

1. Combine the following reagents with the 2 µL of mRNA sample bound to the beads: 1 µL of SMART oligo (10 µ$M$) and 1 µL of Oligo(dT)-T7 promoter primer (10 µ$M$). Mix the contents and spin the tube briefly.
2. Incubate at 72°C for 2 min. Cool on ice for 2 min (*see* **Note 4**).
3. Centrifuge briefly to collect the contents at the bottom.
4. Add the following reagents to each tube: 2 µL of 5X first-strand buffer, 1 µL of DTT (20 m$M$), 1 µL of dNTP mix (10 m$M$), 1 µL of Superscript II (200 U/µL),

and 1 µL of T4 gp32 (150 ng/µL) (*see* **Note 5**). Mix the contents by brief centrifugation for a final volume of 10 µL.

5. Incubate at 42°C for 1.5 h (using a thermal cycler or block).
6. Place the tubes on ice in preparation for the next step or store the samples at –20°C.

### 3.3. cDNA Amplification by Long Distance PCR to Generate T7-Promoter-Linked Double-Stranded cDNA

1. Combine the following reagents with the 10 µL of first-strand cDNA from **step 6** of **Subheading 3.2.**: 10 µL of 10X Advantage 2 PCR buffer, 2 µL of 50X dNTP mix, 2 µL of Oligo(dT)-T7 promoter primer (10 µ*M*), 2 µL of 5' PCR (10 µ*M*), 2 µL of Advantage 2 Polymerase mix (*see* **Note 6**), and 72 µL of nuclease-free water, for a total volume of 100 µL. Mix the contents and centrifuge briefly.
2. Commence cycling using the following program: 95°C for 1 min, then 30 cycles at 95°C for 15 s and 68°C for 5 min The double-stranded cDNA can be processed immediately or stored at –20°C indefinitely (–70°C).

### 3.4. Quality Controls Employing Gene-Specific PCRs and Agarose Gel Electrophoresis

The purpose of this procedure is to monitor the success of the amplification procedure. The tube containing no mRNA template (negative control) usually yields a cDNA smear consisting of primer concatamers.

1. Run 5 µL of each sample on a 2% agarose gel to monitor the cDNA size range as illustrated in **Fig. 1A**.
2. Take a 0.5-µL sample for *β-ACTIN* expression analysis by PCR (*see* **Note 7**): 1.0 µL (25 pmol/µL) of each primer (sense and antisense), 1.0 µL (10 µ*M*) of dNTPs, 2.5 µL (Advantage 2 PCR buffer) of 10X PCR, 0.5 µL of Advantage 2 Polymerase, and 18.5 µL of nuclease-free water.
3. Commence cycling using the following program: 95°C for 5 min, then 25 cycles at 95°C for 30 s, 60°C for 30 s, 72°C for 30 s, and a final extension at 72°C for 10 min.

#### 3.4.1. Gel Electrophoresis

1. After PCR amplification, resolve 5 µL of the reaction product on a 1.0% agarose gel containing 2 µg/mL of ethidium bromide for 60 min at 100 V (**Fig. 1B**).

#### 3.4.2. Purification and Quantitation of cDNAs

1. Use the NucleoSpin Extraction II kit following the recommended protocols.
2. Elute the cDNAs with 20 µL of nuclease-free water, quantitate, and store at –20°C until required (*see* **Note 8**).

### 3.5. Processing of cDNAs for Ligation into Plasmid or Lambda-Based Vectors

1. Digest the cDNAs with *Sfi*I.
2. Enrich for large cDNAs by size fractionation on a low-melting-point agarose gel. Excise cDNAs >1.0 kb.

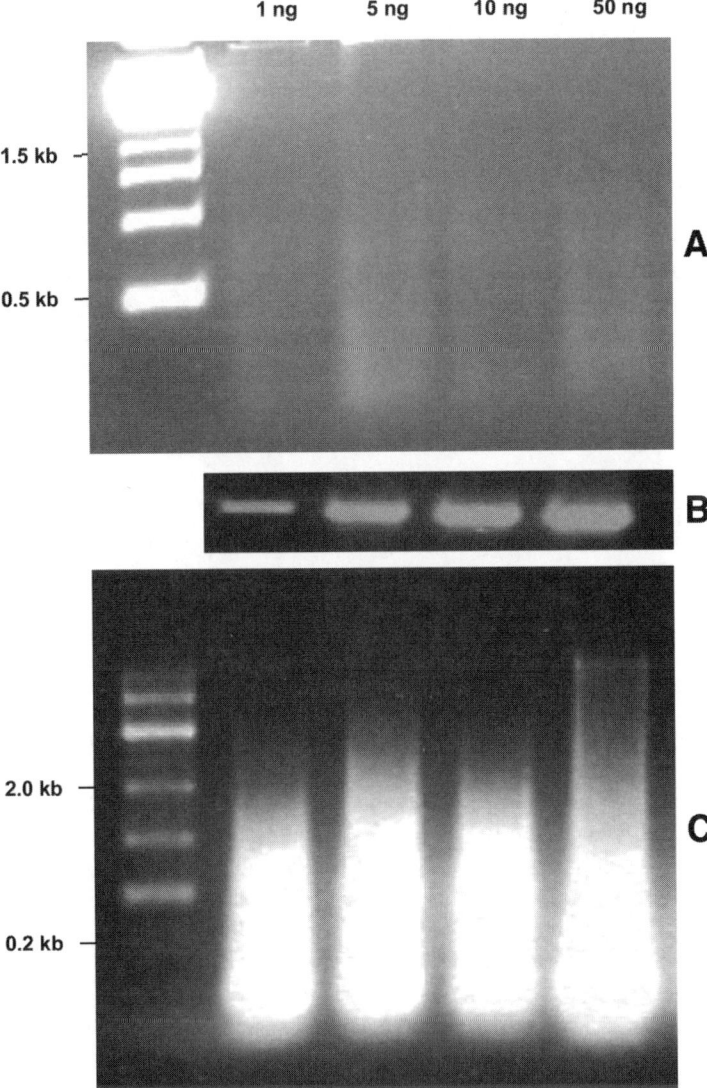

Fig. 1. Qualitative analysis of generated cDNAs and aRNAs. (**A**) The generated cDNAs are in a size range of >1.5 kb. (**B**) The expression of *β-ACTIN* in all samples is shown. The gradual increase in concentration of the PCR products reflects the increasing concentration of cDNAs generated with increasing concentration of input mRNA. (**C**) Corresponding aRNAs have a size distribution of >2.0 kb in length. However, the bulk of the aRNAs are in the range of up to 2.0 kb.

3. Elute the excised cDNAs using the NucleoSpin Extraction II kit.
4. Directionally clone cDNAs into pTriplEx2 phagemid or pDNR-LIB plasmid, which are optimized for efficient cloning of large inserts. Follow the manufacturer's

**Table 1**
**Yields of Double-Stranded cDNAs and Corresponding aRNAs[a]**

| HeLa total RNA (ng) | Total cDNA (µg) | aRNA (µg/µL) per 0.5 µg input cDNA | Total amount of aRNA (µg) generated per 0.5 µg input cDNA |
|---|---|---|---|
| 1[b] | 4.65 | 5.9 | 118 |
| 5 | 5.43 | 5.6 | 112 |
| 10 | 6.16 | 5.8 | 116 |
| 50 | 7.0 | 6.3 | 126 |

[a]The concentrations of total RNAs used (1, 5, 10, and 50 ng) are equivalent to 0.05, 0.25, 0.5, and 2.5 ng of mRNA, respectively, assuming that the composition of mRNA within a cell is 5% of the total RNA. In all cases, the mRNA amplification is more than a million-fold.

[b*]One nanogram of total RNA is equivalent to about 25 cells.

recommended protocols for ligation, transformation, propagation, and storage of cloned libraries.

### 3.6. Generation of Amplified RNA Suitable for Microarray-Based Expression Profiling

1. Use 500 ng of input double-stranded cDNA as template for the in vitro transcription reaction using the MessageAmp aRNA kit adhering to the recommended protocols.
2. Allow the reaction to proceed for 5 h.
3. Add 2 µL (4 U) of DNase1 and incubate for 30 min at 37°C. Follow the manufacturer's recommended protocol for purification.
4. Elute aRNA with 50 µL of elution buffer. Evaluate the purity and concentrations using an Agilent 2100 Bioanalyzer or a suitable spectrophotometer (*see* **Table 1** and **Note 9**). Alternatively, check the size range and integrity of the aRNA by resolving 0.5 µL on a denaturing or nondenaturing agarose gel employing standard procedures for electrophoresis (*see* **Fig. 1C**).
5. Monitor the quality of the aRNA by carrying out gene-specific RT-PCRs (optional). The RNA should be stored at –70°C. The quality of the RNA is good enough for use in microarray expression profiling experiments.

## 4. Notes

1. The primers used for this protocol contain sequences for both the T7 promoter recognition site and the restriction sites for *Sfi* 1A and 1B. Cloning of cDNAs using these restriction sites will result in cDNAs ligated in a 5' to 3' direction. This is a useful feature if expression libraries are required. Other restriction sites can also be designed to replace *Sfi* 1A and 1B.
2. If starting with previously purified total RNA, ensure that the starting volume does not exceed 5 µL.

3. Extreme care should be taken when washing the beads. After the brief spin, the magnetic beads will be visible as a tiny brown pellet at the bottom of the tube. Avoid long delays between washing steps. If working with multiple samples, use a multiple-station magnetic particle concentrator.

4. To avoid evaporation and condensation during the 72°C incubation, overlay a single drop of mineral oil to cover the reaction contents.

5. To augment the loss of 5' complexity, include T4 single-strand nucleic acid binding protein (T4gp32) in all reverse transcription reactions. The combination of this enzyme in the reverse transcription reaction increases the yield of RT-PCR products by as much as 50% and presumably makes the reaction more robust and therefore reproducible. The mechanism of action of T4 protein is poorly understood. However, it is known that it can interact with DNA and proteins at the replication fork, thus increasing the efficiency and fidelity of DNA replication *(14–15)*. The protein is also suspected to interact with RNA and influence mRNA or RNA/DNA duplexes during reverse transcription.

6. The Advantage polymerase system consists of two different DNA polymerases (primary and proofreading), thus allowing the amplification of significantly longer fragments in a process referred to as long distance PCR *(16)*. Other combinations of polymerases can be used to achieve the same effect.

7. The housekeeping gene *b-ACTIN* has been used for quality control purposes, but this can be augmented with additional control genes relevant to the cell type under investigation, such as the expression of *OCT4* in the case of embryonic cells *(3)*.

8. If the sole purpose of the experiment is to generate amplified RNA, then it would be better to concentrate the eluted cDNA by partial drying down in a SpeedVac to increase the concentration to between 200 and 500 ng/µL. This level of concentration is optimal for in vitro transcription.

9. If the final concentration of the aRNA generated is below that required for subsequent experiments, a second round of linear amplification can be carried out using the first-round aRNA as template. All necessary reaction components are supplied with the MessageAmp aRNA kit.

## Acknowledgments

I thank Prof. Hans Lehrach for his support. This work was funded by the German Ministry for Education and Research (BMBF) as part of the National Genome Research Network (NGFN).

## References

1. Roozemond, R. C. (1976) Ultramicrochemical determination of nucleic acids in individual cells using the Zeiss UMSP-I microspectrophotometer: application to isolated rat hepatocytes of different ploidy classes. *Histochem. J.* **8,** 625–638.

2. Phillips, J. and Eberwine, J. H. (1996) Antisense RNA amplification: a linear amplification method for analyzing the mRNA population from single living cells. *Methods* **10,** 283–288.

3. Adjaye, J., Bolton, V., and Monk, M. (1999) Developmental expression of specific genes detected in high quality cDNA libraries from single human preimplantation embryos. *Gene* **237,** 373–383.

4. Goto, T., Adjaye, J., Rodeck, C., and Monk, M. (1999) Identification of human primordial germ cell specific transcribed genes by differential display. *Mol. Hum. Reprod.* **5,** 851–860.

5. Adjaye, J. and Monk, M. (2000) Transcription of homeobox-containing genes in cDNA libraries derived from human oocytes and preimplantation embryos. *Mol. Hum. Reprod.* **6,** 707–711.

6. Smelt, V. A., Upton, A., Adjaye, J., et al. (2000) Expression of arylamine N-acetyltransferases in pre-term placentas and in human pre-implantation embryos. *Hum. Mol. Genet.* **9,** 1101–1107.

7. Pitera, J., Milla, P., Scambler, P., and Adjaye, J. (2001) Cloning of *HOXD1* from unfertilised human oocytes and expression analyses during murine oogenesis and embryogenesis. *Mech. Dev.* **109,** 377–381.

8. Ponsuksili, S., Wimmers, K., Adjaye, J., and Schellander, K. (2002) A source for expression profiling in single preimplantation bovine embryos. *Theriogenology* **57,** 1611–1624.

9. Zhumabayeva, B., Chenchik, A., Siebert, P. D., and Herrler, M. (2004) Disease profiling arrays: reverse format cDNA arrays complimentary to microarrays. *Adv. Biochem. Eng. Biotechnol.* **86,** 191–213.

10. Wellenreuther, R., Schupp, I., The German cDNA Consortium, Poustka, A., and Wiemann, S. (2004) SMART amplification combined with cDNA size fractionation in order to obtain large full-length clones. *BMC Genomics* **5,** 36.

11. Wang, E., Miller, L. D., Ohnmacht, G. A., Liu, E. T., and Marincola, F. M. (2000) High-fidelity mRNA amplification for gene profiling. *Nat. Biotechnol.* **18,** 457–459.

12. Wang, Q. T., Piotrowska, K., Ciemerych, M. A., et al. (2004) Genome-wide study of gene activity reveals developmental signaling pathways in the preimplantation mouse embryo. *Dev. Cell* **6,** 133–144.

13. Adjaye, J. (2005) Whole-genome approaches for large-scale gene identification and expression analysis in mammalian preimplantation embryos. *Reprod. Fertil. Dev.* **17,** 37–45.

14. Frohman, M. A., Dush, M. K., and Martin, G. R. (1988) Rapid production of full-length cDNAs from rare transcripts: amplification using a single gene-specific oligonucleotide primer. *Proc. Natl. Acad. Sci. USA* **85,** 8998–9002.

15. Chandler, D. P., Wagon, C. A., and Bolton, H. (1998) Reverse transcription inhibition of PCR at low concentrations of template and its implications for quantitative RT-PCR. *Appl. Environ. Microbiol.* **64,** 669–677.

16. Barnes, W. M. (1994) PCR amplification of up to 35-kb DNA with high fidelity and high yield from λ-bacteriophage templates. *Proc. Natl. Acad. Sci. USA* **91,** 2216–2220.

# 11

## Use of Real-Time Polymerase Chain Reaction to Measure Gene Expression in Single Cells

### Dagan Wells

### Summary

Quantification of the expression of individual genes can reveal much concerning the processes occurring within a cell. In the vast majority of cases, activation or repression of a gene is indicative of altered utilization of the pathway or process in which it functions. Although microarray analysis has the power to provide data concerning the expression of thousands of genes in a single experiment, validation of results using alternative methods is still essential. This is particularly true if the amount of RNA available for microarray analysis is very small, necessitating methods of RNA amplification. The gold-standard for quantifying mRNA transcripts from an individual gene is the use of reverse transcription followed by real-time PCR. This approach has yielded highly accurate and reproducible data, even when applied to minute samples, such as single oocytes or single embryos. This chapter describes protocols for the quanitification of mRNA transcripts using real-time PCR and considers issues specific to analysis of single cells.

**Key Words:** Single cell; oocyte; embryo; blastomere; real-time PCR; reverse transcription; RNA; gene expression.

## 1. Introduction

The study of gene expression provides many important clues concerning the temporal, spatial, and developmental regulation of biochemical pathways within cells. Key processes such as metabolism, differentiation, and apoptosis are all influenced by fluctuations in the activity of specific genes. Not only does the analysis of gene expression improve the understanding of fundamental cellular mechanisms, but it is also likely to have far-reaching clinical implications in a number of fields, including oncology and assisted reproductive technology, where it may assist in the identification of novel diagnostic and therapeutic targets.

In the field of assisted reproductive technology, much effort is currently directed toward optimization of in vitro culture methods for human embryos

From: *Methods in Molecular Medicine: Single Cell Diagnostics: Methods and Protocols*
Edited by: A. Thornhill © Humana Press Inc., Totowa, NJ

and techniques for identifying the embryos that have the greatest probability of forming a successful pregnancy. In both these cases analysis of gene expression is expected to lead to substantial improvements in clinical methodology. The majority of human preimplantation embryos generated by in vitro fertilization fail to progress beyond the preimplantation phase, degenerating or becoming irreversibly arrested. The failure of such embryos contributes to the low (~20%) implantation rate achieved during in vitro fertilization (IVF) treatment *(1,2)*.

The preimplantation stage is one of the most challenging phases of mammalian development. During this time an embryo must initiate several fundamental processes, including genome activation, compaction, cavitation, and the first cellular differentiation. Currently, the mechanisms underlying the regulation of these processes are poorly understood. However, it is likely that carefully controlled changes in gene expression are central to the successful navigation of preimplantation development. Variation in gene activity is largely responsible for the subtle control of cellular and biochemical pathways and has been shown to have a profound impact on tissue growth and architecture during the development of a wide variety of organisms.

In an effort to improve the success rates of IVF, attempts are being made to identify and preferentially transfer the most viable embryos from each cohort. Traditionally, methods for assessing embryo viability have been based on morphological evaluation (e.g., *see* **refs. *3–7*)**. However, approaches of this type have only a limited power to distinguish viable embryos from those that are compromised. It is now becoming increasingly common for morphological examination to be complemented with chromosomal ploidy information obtained using preimplantation genetic diagnosis (PGD) *(8–10)*. PGD usually involves the culture of embryos for 3 d postfertilization, by which time most embryos are composed of 6–10 cells. At this point a single cell can be biopsied from the embryo and subjected to genetic testing. The use of PGD methods for assessing chromosomes allows physicians to distinguish chromosomally normal embryos from those that harbor lethal chromosome abnormalities—a common occurrence in human embryos. Chromosomally normal embryos are considered to have a higher probability of implanting and forming a pregnancy than their aneuploid counterparts and, consequently, it is advantageous to ensure that these embryos are preferentially transferred to the mother.

Chromosomal screening of preimplantation embryos increases IVF implantation rates for certain groups of infertile patients (*see* **ref. *11*** for a review). However, chromosomal imbalance can only partially account for the high incidence of embryonic arrest seen in human IVF. It is possible that more subtle forms of preimplantation analysis focused on determining the activity of specific genes or cellular pathways could provide a more precise indication of embryo viability. Particular patterns of expression might be expected in embryos suffering thermal shock, embryos with abnormal chromosome numbers, or those in the process of undergoing developmental arrest. Similarly, assessing the activity of pathways associated with specific forms of cellular stress may assist in the identification of specific deficiencies in embryo culture media.

Before the analysis of gene expression can be considered a feasible method for measuring embryo viability, it will be necessary to define patterns of gene expression that are characteristic of healthy, morphologically normal embryos. It will also be important to determine whether specific abnormalities result in identifiable changes in gene expression characteristics. We have recently shown that some abnormal embryo morphologies do indeed result in disturbances of normal patterns of gene expression *(12,13)*.

A major consideration when designing tests for measuring gene expression in human embryos is that current embryo biopsy methods allow only a single cell to be sampled. Thus, any gene expression–based PGD methods must be applicable at the single cell level. Applicability to low quantities of RNA is also essential for most gene expression analysis of human embryos carried out in a research context, owing to the extremely limited availability of embryos donated for research.

Low quantities of sample material can be problematic in areas of clinically oriented research aimed at investigating small numbers of tumor cells isolated from the bloodstream, feces, or urine. Such tests hold the promise of noninvasive methods for routine cancer screening but require much greater sensitivity than is provided by most forms of genetic analysis. Minimally invasive forms of prenatal diagnosis that assess fetal cells isolated from maternal blood or the cervical canal also focus on low cell numbers or even single cells. Thus, it is likely that the development of sensitive tests for the analysis of gene expression in minute samples will be of scientific and clinical value in a number of diverse fields. This chapter details protocols that permit accurate quantification of mRNA transcript numbers in extremely limited tissue samples, including single cells.

## 2. Materials

1. DNA-Zap (Ambion, Austin, TX).
2. RNA-Zap (Ambion).
3. Acidified Tyrode's solution (Sigma, St. Louis, MO).
4. Phosphate-buffered saline (PBS) (Invitrogen, Carlsbad, CA).
5. Polyvinyl alcohol (PVA) (Sigma).
6. RNasin (Promega, Madison, WI).
7. Micro RNA Isolation kit (Stratagene, La Jolla, CA).
8. GeneAmp RNA PCR kit (Applied Biosystems, Foster City, CA).
9. Microcentrifuge.
10. Glycogen (Roche, Indianapolis, IN).
11. Diethylpyrocarbonate (DEPC)-treated water (Ambion).

12. Amplification Grade DNase I (Invitrogen).
13. Agarose (Invitrogen).
14. TBE buffer (Invitrogen).
15. Ethidium bromide (Bio-Rad, Hercules, CA).
16. Electrophoresis equipment.
17. MinElute Gel Extraction Kit (Qiagen, Valencia, CA).
18. Spectrophotometer.
19. 10X Polymerase chain reaction (PCR) buffer (30 m$M$ $MgCl_2$) (Idaho Technology, Salt Lake City, UT).
20. Oligonucleotide primers.
21. Deoxynucleotides (Promega).
22. SYBR green DNA stain (Molecular Probes, Eugene, OR).
23. Taq polymerase (Applied Biosystems).
24. TaqStart antibody (Clontech, Palo Alto, CA).
25. LightCycler™ real-time PCR machine (Roche).

## 3. Methods

### 3.1. Samples

The limited availability of human embryos and oocytes for research means that, by necessity, genetic analysis must focus on extremely small amounts of nucleic acids (i.e., the DNA or RNA from single oocytes or embryos). There are few disciplines that deal with samples on a similarly restricted scale, notable examples being forensics, molecular archaeology, and certain experimental forms of tumor detection or noninvasive prenatal diagnosis. The analysis of minute quantities of nucleic acid exacerbates problems encountered in other areas of genetic research, as well as introduces difficulties that are unique to the investigation of very small samples, such as single cells.

### 3.1.1. General Precautions Against Contamination

Contamination with extraneous DNA fragments, an easily manageable problem for most laboratories undertaking genetic research, represents a major difficulty for laboratories performing reverse transcriptase (RT)-PCR analyses of embryos, blastomeres, polar bodies, oocytes, and other single cells and must be avoided at all costs. General precautions against contamination include wearing gloves, using meticulously cleaned equipment dedicated to the handling of single cells, and performing sample preparation and PCR setup in a hood. If possible, the hood should be irradiated with ultraviolet light for 30 min prior to each use and cleaned with reagents that degrade DNA and/or RNA (e.g., DNA-Zap and RNA-Zap; Ambion). If DNA amplification is being carried out, it is important to have a physical separation between the area used for PCR setup and the area in which amplified DNA is analyzed or manipulated. Ideally, amplification reactions should be set up in a separate room with limited access by personnel. Equipment and lab coats should not be allowed to pass from the analytical laboratory to the PCR setup room without first undergoing a rigorous decontamination. Quantitative real-time PCR, which is frequently used to assess the copy number of specific mRNA transcripts, usually involves the amplification of previously amplified DNA fragments of known concentration (*see* **Subheading 3.4.**). Such fragments pose a serious contamination risk to the samples being simultaneously assessed and should not be introduced in the room used for PCR setup.

### 3.1.2. Preparation of Sample

As well as the sources of contamination already discussed, oocyte and embryo samples can be contaminated with nucleic acids derived from spermatozoa or cumulus cells attached to the zona pellucida. For this reason, our previous experiments have involved removal of the zona pellucida using a brief immersion in acidified Tyrode's solution (Sigma). Ooctyes/embryos were then washed through three droplets of PBS supplemented with 0.1% (w/v) PVA and 0.3 U/μL of RNasin (Promega). The PVA helps to prevent the cells from sticking, while the RNasin protects the RNA from enzymatic degradation. Removal of the zona pellucida and washing should be performed rapidly (*see* **Note 1**). After washing, samples were transferred to a microfuge tube in ~1 μL of PBS/PVA/RNasin to which 100 μL of denaturing solution and 0.72 μL of β-mercaptoethanol (Micro RNA Isolation kit; Stratagene) was added. Samples were stored immediately at –80°C. After addition of the denaturing solution, samples could be stored at –80°C for more than 6 mo.

## 3.2. RNA Extraction

### 3.2.1. Precautions and Controls

Throughout the RNA extraction procedure every effort was made to maintain an aseptic technique: Manipulations were carried out in a hood that had been cleaned with RNA-Zap (Ambion,; the pipets used were designated for RNA use only, sterile tubes and pipet tips were utilized, and gloves were worn at all times. Prior to RNA extraction each sample was spiked with $10^6$ copies of an RNA transcript derived from a plasmid (pw109) (GeneAmp RNA PCR kit; Applied Biosystems). The quantity of this nonhuman RNA was assessed after RNA extraction to provide an indication of extraction efficiency *(14)*. Samples displaying less plasmid RNA than controls were excluded from further study.

### 3.2.2. Purification of RNA

1. To each sample add 10 μL of 2 *M* sodium acetate (pH 4.0), 100 μL of water-saturated acid phenol, and 30 μL of chloroform:isoamyl alcohol (Micro RNA Isolation kit; Stratagene).

2. Mix the samples thoroughly and centrifuge for 5 min at 16,000g.
3. Transfer the aqueous layers to fresh 0.5-mL tubes (DNase and RNase free, Eppendorf), and add 1 µL of glycogen (20 mg/mL; Roche) and 100 µL of isopropanol.
4. Mix the solutions thoroughly and centrifuge for 45 min at 16,000g.
5. Wash the resultant pellet with 200 µL of 75% ethanol and centrifuge for 5 min at 16,000g before air-drying the pellet and resuspending in 5.8 µL of DEPC-treated water (Ambion).
6. To eliminate residual genomic DNA from the RNA sample, add 0.725 µL of 10X Amplification Grade DNase I buffer and 0.725 U of DNase I (Invitrogen) to the sample and incubate at room temperature for 15 min.
7. Halt DNA digestion by adding 1 µL of 25 mM EDTA and heating for 10 min at 65°C.

### 3.3. Reverse Transcription

1. To each RNA sample add the following: 0.2 µL of dithiothreitol (0.1 M), 1.5 µL of oligo dT (50 µM) (RNA PCR kit; Applied Biosystems), 1.05 µL of RNase inhibitor (20 U/µL; RNA PCR kit), 4 µL of 25 mM MgCl₂, 2 µL of 10X PCR buffer (RNA PCR kit), and 4 µL of dNTPs (2 mM dATP, dCTP, dGTP, dTTP).
2. Heat the mixture to 70°C for 6 min and place on ice.
3. Add 0.5 µL of MMLV RT (Perkin Elmer RNA PCR kit), and incubate at 37°C for 1 h followed by 95°C for 5 min. The resultant cDNA samples can be stored at –80°C and are considered stable for at least 1 yr provided that they are not subjected to repeated freeze-thaw cycles (*see* **Note 2**).

### 3.4. Real-Time PCR

### 3.4.1. Controls and Standards for Real-Time PCR

Each sample was tested three times for each gene in order to control for problems such as imperfections in the PCR tube/capillary. The amplification of samples was conducted simultaneously with amplification of four different concentration standards (also run in triplicate) and a negative control composed of reaction mixture with no cDNA added. Each concentration standard contained a known number of copies of the cDNA fragment of interest. The standards were created as follows: DNA fragments from the gene of interest were amplified from a cDNA sample using essentially the same PCR conditions as used for real-time PCR. After amplification was completed, the entirety of the PCR product was loaded into a 2% agarose/1X TBE gel stained with ethidium bromide (Bio-Rad) and subjected to electrophoresis at 50 V for ~30 min (the precise duration of electrophoresis depended on the size of the amplified fragment). Then the band containing the amplified DNA fragment was excised from the gel, and DNA was extracted using a MinElute Gel Extraction Kit (Qiagen) according to the manufacturer's instructions. Next the concentration of the purified DNA was assessed by spectrophotometry (calculated from the absorbance at 260 nm). The molecular weight of the amplicon was then calculated based on the number of occurrences of each type of nucleotide in the DNA fragment and its respective molecular weight. Finally, the number of DNA fragments per microliter was calculated based on the concentration of the DNA and the molecular weight of the fragments. For example, the measured concentration of DNA extracted from the gel = 1.6 μg/mL, and the molecular weight of the DNA fragment = 46,281. Therefore, 1 mL of the extracted DNA solution contains $3.457 \times 10^{-11}$ mol of fragments, i.e., $1.6 \times 10^6/46,281$. This number of fragments contained in 1 mL is equal to Avogadro's number ($6.022 \times 10^{23}$) multiplied by $3.457 \times 10^{-11} = 2.082 \times 10^{13}$.

### 3.4.2. Real-Time Quantitative PCR

In the protocol described here the amount of amplified DNA was monitored by SYBR green, a molecule that fluoresces in the presence of double-stranded DNA: the more amplified the DNA the greater the fluorescence. Alternative options for assessing accumulation of PCR products are also available, such as TaqMan probes and molecular beacons, both of which employ probes that fluoresce in the presence of a specific DNA sequence. SYBR green has the advantage that it can interact with any DNA sequence and, consequently, it can be used for the purposes of quantification in any amplification reaction but should be applied only in well-optimized amplification reactions (*see* **Note 3**).

1. For reactions of 30-μL volume, use 2 μL of cDNA, 3 μL of 10X buffer (30 m$M$ MgCl$_2$; Idaho Technology), 0.5 μ$M$ of each gene-specific primer, 0.2 m$M$ dNTPs (Promega), SYBR green DNA stain (1 μL of a 1/4000 dilution of concentrated stock; Molecular Probes), and 1.88 U of Taq polymerase (inactivated by the addition of TaqStart antibody; Clontech).
2. Add 7 μL of this mixture to each of three capillary tubes, seal the tubes, and subject to thermal cycling using a LightCycler real-time PCR machine (Roche).
3. Carry out amplification using a program consisting of heating at 96°C for 1 min, followed by 40–50 cycles of 0 s at 95°C, 50–60°C (depending on primers) for 0 s, and 72°C for 10–15 s (*see* **Note 4**).
4. Acquire fluorescence data during an additional step at approx 3°C below the product $T_m$ for 2 s.
5. Confirm product identity by ethidium bromide–stained 2% agarose gel electrophoresis and/or DNA sequencing.

### 3.5. Quantification

Data were analyzed using software supplied with the LightCycler. A standard curve was generated by reference to standards containing a known quantity of amplicons. Fluorescence was acquired at the end of each cycle in order to determine the cycle during the log-linear phase of the reaction at which fluorescence rises above background for each sample (i.e., amplified product becomes detectable) (*see* **Fig. 1**). The LightCycler quantification software generates a

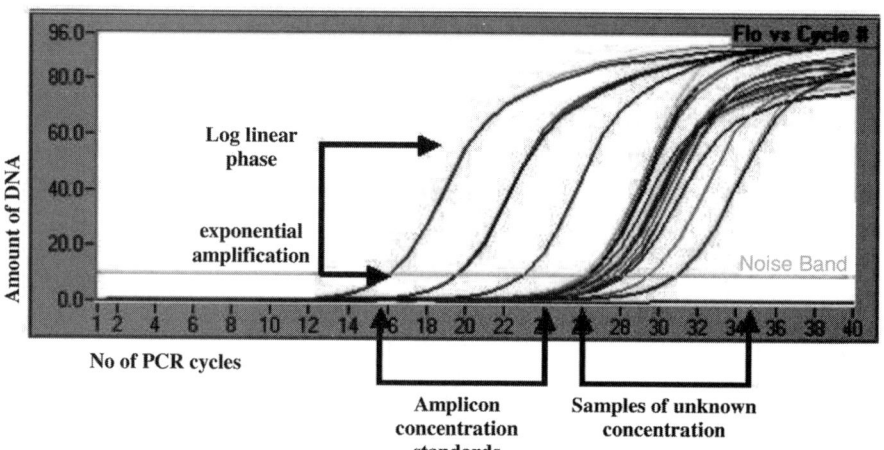

Fig. 1. Real-time PCR after reverse transcription in single cells. The accumulation of amplified DNA is measured in each sample tube in real time (i.e., once per cycle). All samples are run in triplicate and assessed relative to standards containing a known number of template molecules. If measurements are taken after fluorescence rises above the background and while the amplification is still proceeding in an exponential fashion, it is possible to calculate the number of templates in each sample by reference to the controls.

best-fit line and determines unknown concentrations by interpolating the noise-band intercept of an unknown sample against the standard curve of known concentrations *(14,15)*.

## 4. Notes

1. Any handling of cells prior to placement in denaturation buffer should be as brief as possible, because environmental changes (alteration in pH, media composition, and temperature) are likely to elicit changes in gene expression. Generally, the steps described require <1 min/sample.
2. Cycles of freezing and thawing lead to degradation of nucleic acids. Consequently, freeze-thawing of samples should be minimized. Additionally, concentration standards should be divided into single-use aliquots and frozen at –80°C as soon as they have been produced. Aliquots should be discarded after thawing.
3. SYBR green is unable to distinguish between amplification of the intended fragment and spurious PCR products (e.g., primer-dimers and nonspecific PCR products) and, as a consequence, is better suited to well-optimized reactions in which nonspecific products are rare.
4. Alternative real-time PCR machines other than the LightCycler generally require longer incubations at denaturing and annealing temperatures.

## References

1. Ebner, T., Yaman, C., Moser, M., Sommergruber, M., Polz, W., and Tews, G. (2001) Embryo fragmentation in vitro and its impact on treatment and pregnancy outcome. *Fertil. Steril.* **76,** 281–285.
2. 2001 Assisted Reproductive Technology Success Rates; National Summary and Fertility Clinic Reports (2003), Atlanta: Center for Disease Control and Prevention, p. 7 (http://www.cdc.gov/reproductivehealth/ART00/index.htm).
3. Giorgetti, C., Terriou, P., Auquier, P., Hans, E., Spach, J. L., Salzmann, J., and Roulier R. (1995) Embryo score to predict implantation after in-vitro fertilization: based on 957 single embryo transfers. *Hum. Reprod.* **10,** 2427–2431.
4. Ziebe, S., Petersen, K., Lindenberg, S., Andersen, A. G., Gabrielsen, A., and Andersen, A. N. (1997) Embryo morphology or cleavage stage: how to select the best embryos for transfer after in-vitro fertilization. *Hum. Reprod.* **12,** 1545–1549.
5. Rijinders, P. M. and Jansen, C. A. M. (1998) The predictive value of day 3 embryo morphology regarding blastocyst formation, pregnancy, and implantation rate after day 5 transfer following in vitro fertilization or intracytoplasmic sperm injection. *Hum. Reprod.* **13,** 2869–2873.
6. Van Royen, E., Mangelschots, K., De Neubourg, D., et al. (1999) Characterization of a top quality embryo, a step towards single-embryo transfer. *Hum. Reprod.* **14,** 2345–2349.
7. Milki, A. A., Hinckley, M. D., Gebhardt, J., Dasig, D., Westphal, L. M., and Behr, B. (2002) Accuracy of day 3 criteria for selecting the best embryos. *Fertil. Steril.* **77,** 1191–1195.
8. Gianaroli, L., Magli, M. C., Ferraretti, A. P., Tabanelli, C., Trombetta, C., and Boudjema, E. (2002) The role of preimplantation diagnosis for aneuploidies. *Reprod. Biomed. Online* **4(Suppl. 3),** 31–36.
9. Munne, S., Sandalinas, M., Escudero, T., et al. (2003) Improved implantation after preimplantation genetic diagnosis of aneuploidy. *Reprod. Biomed. Online* **7,** 91–97.
10. Verlinsky, Y. and Kuliev, A. (2004) Preimplantation diagnosis for aneuploidies in assisted reproduction. *Minerva Ginecol.* **56,** 197–203.
11. Munne, S. and Wells, D. (2002) Preimplantation genetic diagnosis. *Curr. Opin. Obstet. Gynecol.* **14,** 239–244.
12. Wells, D., Bermudez, M. G., Steuerwald, N., Thornhill, A. R., Malter, H., Delhanty, J. D. A., and Cohen, J. (2005) Expression of genes regulating chromosome segregation, the cell cycle and apoptosis during human preimplantation development. *Hum. Reprod.* **20,** 1339–1348.
13. Wells, D., Bermudez, M. G., Steuerwald, N., Malter, H., Thornhill, A. R., and Cohen, J. (2005) Association of abnormal morphology and altered gene expression in human preimplantation embryos. *Fertil. Steril.* **84(2),** 343–355.
14. Steuerwald, N., Cohen, J., Herrera, R. J., and Brenner, C. A. (2000) Quantification of mRNA in single oocytes and embryos by real-time rapid cycle fluorescence monitored RT-PCR. *Mol. Hum. Reprod.* **6,** 448–453.
15. Wittwer, C. T., Herrmann, M. G., Moss, A. A., and Rasmussen, R. P. (1997) Continuous fluorescence monitoring of rapid cycle DNA amplification. *BioTechniques* **22,** 130–138.

# 12

## Gender Determination and Detection of Aneuploidy in Single Cells Using DNA Array–Based Comparative Genomic Hybridization

**Dong Gui Hu, Xin Yuan Guan, and Nicole Hussey**

### Summary

Comparative Genomic Hybridization (CGH) using metaphase chromosome spreads to screen all human chromosomes for aneuploidy in preimplantation embryos is hindered by the time required to perform the analysis. DNA microarrays manufactured to date are not able to analyze the very limited amount of genetic material in a single cell. We have developed a DNA Microarray–Comparative Genomic Hybrydization (CGH) approach, which is capable of screening all 24 human chromosomes for aneuploidy and sex chromosomal abnormalities in a single cell. This technique has a number of applications including aneuploidy detection in single fetal cells isolated noninvasively from pregnant women for prenatal diagnosis. In this chapter, we describe in detail the methodology of this technique, including preparation of a single cell sample, whole genome amplification of single cells, microarray CGH hybridization (array CGH), data analysis of array CGH, and the diagnostic criteria of aneuploidy detection. We anticipate that with further automation of part of this technique it will become a high-throughput diagnostic tool.

**Key Words:** Microarray CGH; aneuplody, preimplantation genetic diagnosis, early noninvasive prenatal diagnosis, single cell diagnostics.

## 1. Introduction

Chromosomal abnormalities, such as aneuploidy, occur frequently in human preimplantation embryos and may be of increased frequency in those created by in vitro fertilization, because rates up to 75% abnormal embryos have been reported using fluorescence *in situ* hybridization (FISH) *(1–4)*. Over the last decade, FISH has been utilized to screen embryos for aneuploidy of a limited number of chromosomes via a procedure termed preimplantation genetic diagnosis (PGD) *(3,5)*. Typically, this screening covers only chromosomes 13, 14, 15, 16, 18, 21, 22, X, and Y whereas meiotic and mitotic errors found in preimplantation embryos involve almost every chromosome *(1,6–8)*. Comparative

From: *Methods in Molecular Medicine: Single Cell Diagnostics: Methods and Protocols*
Edited by: A. Thornhill © Humana Press Inc., Totowa, NJ

genomic hybridization (CGH) is a technique that compares the amount of genetic material present in one sample with that of a known normal standard. The technique involves amplifying the genetic material in the unknown sample (e.g., a single blastomere) and subsequent labeling with a first fluorescent color and mixing it with amplified, labeled (in a second fluorescent color) normal standard. The mixture is then hybridized to a normal metaphase chromosome spread in the case of metaphase CGH and to a microarray in the case of array CGH. The advantage of CGH over FISH is that it does not require "spreading" of the unknown single blastomere DNA, a step that can lead to lost DNA, over-lapping, and/ or split signals.

Metaphase CGH appears to be reliable for detecting aneuploidy in a single cell sample *(9)*. However, the application of single cell metaphase CGH to screen all chromosomes for aneuploidy in preimplantation embryos is hindered by the long length of time required for the hybridization step alone (up to 72 h) *(9,10)*. Embryos need to be transferred by d 5 or 6 after the biopsy procedure performed on d 3 *(11)*. Array CGH has been widely used to detect copy-num-ber changes of genomic DNA sequences, including heterozygous and homozy-gous deletions of specific genes as well as partial or whole chromosome aneuploidy *(12–15)*. We have developed a DNA microarray containing probes for all 24 human chromosomes and demonstrated its ability to detect aneu-ploidy in a single cell as the starting material within a time frame relevant to PGD (approx 30 h). The Y probe used for the arrays was made by flow sorting and produced the expected results for the Y in about 20% of hybridizations *(16)*. Subsequent FISH analysis showed a weak cross-hybridization to chromosome 21 (unpublished data). We have generated three new Y chromosome probes by microdissection and used them to replace the previous Y probe. In this chapter, we describe in detail single cell array CGH analysis.

## 2. Materials

### 2.1. Sources of Single Cells

Peripheral blood samples of normal male (46,XY) and normal female (46,XX).
Fibroblast cell lines with known trisomies (e.g., 47,XY,+18) purchased from Coriell
   Cell Repositories, Camden, NJ (http://www.coriell.org).

### 2.2. Cell Culture

1. 1X Minimum essential medium (MEM) (Invitrogen).
2. 1X Trypsin-EDTA: 0.25% trypsin + 1 m*M* EDTA.4Na (Invitrogen).
3. 1X Phosphate-buffered saline (PBS), pH 7.2 (Invitrogen).
4. 1X Dulbecco's PBS(Invitrogen).
5. 200 m*M* (100X) L-Glutamine (Invitrogen).
6. Fetal bovine serum (FBS), certified (Invitrogen).
7. Incubator suitable for cell culture.

## 2.3. Preparation of Single Cells

1. Centrifuge (Beckman TJ-6).
2. Lymphoprep™ (Nycomed Pharma AS, Oslo, Norway). DNase I (10 U/µL) (Boehringer Mannheim, GmbH, Germany).
3. Acrodisc Syringe Filters (0.2 µm) (Pall, Ann Arbor, MI).
4. Superfrost microscope glass slides (Menzel-Glaser, Germany).
5. Inverted light microscope (CK2, Olympus, Japan).
6. Sterilized 70% ethanol (Delta West, Perth, WA, Australia).
7. RPMI medium (Sigma).
8. 1X Polymerase chain reaction (PCR) buffer: 50 m$M$ KCl; 10 m$M$ Tris-HCl, pH 8.3.
9. Extruded, cotton-plugged glass Pasteur pipets (Chase, Glens Falls, NY).
10. Thin-walled PCR tubes (0.5 mL) with flat caps (Applied Biosystems, Foster City, CA).

## 2.4. Lysis of Single Cells

1. Lysis buffer: 200 m$M$ KOH, 50 m$M$ dithiothreitol.
2. Neutralization buffer: 300 m$M$ KCl; 900 m$M$ Tris-HCl, pH 8.3; 200 m$M$ HCl.

## 2.5. Single Cell Degenerate Oligonucleotide–Primed PCR

1. AmpliTaq DNA polymerase (Applied Biosystems).
2. 10X PCR buffer: 100 m$M$ Tris-HCl, pH 8.3; 500 m$M$ KCl (Applied Biosystems).
3. 10X K$^+$-free PCR buffer (1 mg/mL of gelatin in 100 m$M$ Tris-HCl).
4. MgCl$_2$ (25 m$M$) (Applied Biosystems).
5. dNTPs (10 m$M$ each of all four dNTPs) (Applied Biosystems).
6. PCR Grade Ultra Pure Water (Biotech International, Perth, WA, Australia).
7. Degenerate oligonucleotide–primed (DOP)-PCR 6MW primer (5'-CCGACTC-GAGNNNNNNATGTGG-3') *(17)*.
8. Minicycler (MJ Research).

## 2.6. Labeling and Purification

1. FluoroLink™Cy3 (Cy3-AP3-dUTP) (Amersham).
2. FluoroLink™Cy5 (Cy3-AP3-dUTP) (Amersham).
3. UltraClean™ PCR Clean-up DNA purification kit (Mo Bio, Sonala Beach, CA).

## 2.7. Agarose Electrophoresis and Digital Photography

1. Standard equipment for agarose mini gel electrophoresis.
2. *SPP*-1 Phage DNA/*Eco*RI (GeneWorks, SA, Adelaide, Australia).
3. *pUC*19 DNA/*Hpa*II (GeneWorks).
4. Kodak digital camera DC120 (Amersham).
5. ID Kodak digital science software (Amersham).

## 2.8. Precipitation of Probes

1. Salmon sperm DNA solution (Gibco-BRL).
2. Human Cot-1 DNA (Gibco-BRL).

3. 3 $M$ Sodium acetate, pH 5.2. Ethanol (99.7–100%) (Merck, Kilsyth, Victoria, Australia).
4. Eppendorf centrifuge at 4°C.
5. Oven at 60°C.
6. PCR machine at 50°C.

## 2.9. Processing Hybridization

1. Hybridization solution: 50% deionized formamide, 0.1% sodium dodecyl sulfate, 5X Denhardt's solution, 3X standard saline citrate (SSC), 10% dextran sulfate.
2. Array slides (*see* **Note 1**) (Reproductive Health Science Pty. Ltd, Department of Obstetrics and Gynaecology, The University of Adelaide, Australia).
3. WT1 slide-warming tray (Ratek, Boronia, VIC, Australia).
4. Hybridization chambers (Corning, Acton, MA).
5. Cell culture incubator at 37°C with a humidity of 95%.
6. Cover slip (10 x 35 mm) (Mediglass, Sydney, NSW, Australia).

## 2.10. Posthybridization Washing

1. Water bath at 45°C.
2. Orbital mixer (Ratek).
3. Microarray high-speed centrifuge (TeleChem).
4. Washing solution: 50% formamide/2X SSC.
5. 2X SSC.
6. 1X SSC.
7. MilliQ water.

## 2.11. Array Scanning and Data Analysis

1. Scanner: GenePix 4000B scanner (Axon, Union City, CA).
2. Software for image analysis: GenePix Pro 4.0.1.17 (Axon).
3. Software for diagnosis: GenoData (Reproductive Health Science Pty. Ltd).

## 3. Methods

### 3.1. Preparation of Lymphocyte Suspensions

1. Transfer 4 mL of a fresh peripheral blood sample into a 10-mL sterilized tube.
2. Centrifuge at 1700 rpm (600$g$) for 10 min in a Beckman TJ-6 centrifuge.
3. Transfer the upper plasma layer into a 1.5-mL Eppendorf tube, and carry out **steps 14–16** as time allows when performing the following steps.
4. Dilute the remaining blood to a volume of about 8 mL with PBS buffer.
5. Layer 2 mL of Lymphoprep solution under the diluted cells to ensure a sharp interface.
6. Centrifuge the tube at 1700 rpm (600$g$) for 20 min.
7. Transfer the lymphocyte layer (the white ring, called a buffy coat) into a fresh 10-mL sterilized tube and dilute to a volume of 10 mL with PBS.
8. Centrifuge the tube at 1700 rpm (600$g$) for 10 min and discard the supernatant.
9. Resuspend the lymphocyte pellet by flicking and add 10 mL of PBS buffer.
10. Centrifuge the tube at 1700 rpm (600$g$) for 10 min and discard the supernatant.

11. Resuspend the lymphocyte pellet by flicking and add 10 mL of PBS buffer.
12. Centrifuge the tube at 1700 rpm (600$g$) for 10 min and discard the supernatant.
13. Resuspend the lymphocyte pellet by flicking in the residual PBS (300–500 µL).
14. Centrifuge the Eppendorf tube containing plasma at 14,000 rpm (10,000$g$) for 20 min, transfer a portion of the upper layer of plasma (so as not to disturb any of the pellet) into a 0.22-µm filter (Millipore), and then filter sterilize into a fresh 1.5-mL sterilized Eppendorf tube.
15. Mix 50 µL of filtered plasma and 2 µL of DNase I (10 U/µL) in a 0.5 mL PCR sterilized tube and incubate the resulting DNase I/plasma mixture at 37°C for 1 h followed by inactivation of the enzyme at 65°C for 10 min. This step was created to remove the genomic DNA probably present in plasma of human peripheral blood.
16. Mix 25 µL of the inactivated DNase I/plasma mixture with 25 µL of lymphocyte suspension from **step 13** to prepare the cell suspension for single cell sorting.

## 3.2. Preparation of Fibroblast Suspensions

Fibroblast cell lines with known chromosomal abnormalities are normally shipped in small flasks containing medium with only 5% FBS and no glutamine, to slow down cell proliferation during transportation.

1. Place newly received flasks in an incubator at 37°C overnight without opening to allow the cells to settle.
2. Discard the shipping medium from the flasks and replace with freshly prepared 1X MEM containing 15% FBS and 2 m$M$ L-glutamine.
3. Harvest the fibroblast cells before they reach confluency using trypsin-EDTA solution.
4. Centrifuge the tube at 1700 rpm (600$g$) for 10 min. Wash the pellet of fibroblast cells using PBS buffer, recentrifuge at 1700 rpm (600$g$) for 10 min to collect the cells, and then resuspend the pellet in the residual PBS (300–500 µL). Cells are now ready for single cell sorting.

## 3.3. Single Cell Sorting

Cell sorting is normally performed on a Superfrost microscope glass slide (Menzel-Glaser) using an inverted light microscope under 200× of magnification (CK2; Olympus); the procedure is depicted in **Fig. 1.**

1. Wash the slide with sterilized 70% ethanol and mount onto the microscope.
2. Pipet 100 µL of RPMI medium onto the left side of the slide to create the RPMI pond.
3. Create three more smaller ponds of approx 50 µL of 1X PCR buffer to the right of the RPMI pond in sequence, and designate them from left to right as PCR buffer pond #1, pond #2, and pond #3, respectively.
4. Add 5 µL of the cell suspension (from **step 16** in **Subheading 3.1.** or **step 4** in **Subheading 3.2.**) to the RPMI pond.
5. Transfer approx 100 cells from the RPMI pond to PCR buffer pond #1 with a 9-in., extruded, cotton-plugged glass Pasteur pipet using a mouth pipet (or equivalent handheld device).

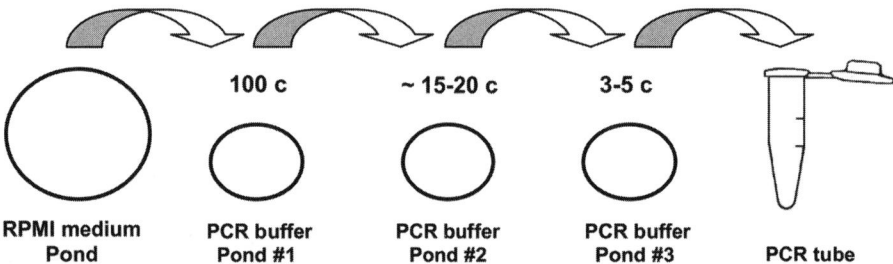

Fig. 1. Single cell sorting procedure. A cell suspension is diluted through four "ponds" placed on a microscope slide to the point where there are only three to five cells in PCR buffer pond #3. A single cell is washed several times in this pond and then transferred into a 0.5-mL PCR tube. PCR buffer (1X) (<3 µL) is aspirated from pond #3 and transferred into a 0.5-mL PCR tube for use as a negative control.

6. Transfer <15–20 cells from PCR buffer pond #1 into PCR buffer pond #2 using a fresh pipet.
7. Using a fresh pipet transfer three to five cells from PCR buffer pond #2 into PCR buffer pond #3.
8. Aspirate one cell from PCR buffer pond #3 using a fresh pipet, and gently move in and out of the pipet at a fresh location in the same pond. After washing, expel this cell (with <3 µL of PCR buffer) into the bottom of a 0.5-mL sterilized PCR tube under microscopic visualization.
9. Isolate more single cells separately from PCR buffer pond #3 using the same pipet.
10. Aspirate a small amount of 1X PCR buffer from the area that had contained the cells in pond #3, and then transfer into a 0.5-mL sterilized PCR tube as a negative control. Tubes containing single cells can be frozen for later use (*see* **Note 2**).

### 3.4. Lysis of Single Cells

1. Add 5 µL of lysis buffer to the 0.5-mL PCR tube containing a single cell.
2. Incubate the tube in a PCR machine at 65°C for 6–10 min.
3. Add 5 µL of neutralization solution to the tube.
4. Spin briefly to collect the lysed solution at the bottom of the tube.

### 3.5. First-Round of DOP-PCR for Random Amplification of Single Cells

1. To the tube containing the lysed and neutralized single cell solution (10 µL), add 5 µL of each of $K^+$-free PCR buffer, $MgCl_2$ (25 m$M$), dNTPs (2.5 m$M$ each of all four dNTPs), and DOP-PCR 6MW primer (20 µ$M$); 1 µL of *Taq* polymerase (5 U/µL); and 19 µL of Ultrapure $H_2O$.
2. Briefly centrifuge the mixture; place in a Minicycler (MJ Research); denature at 95°C for 5 min; and cycle for eight cycles of 94°C for 1 min, 30°C for 1.5 min, 72°C for 3 min with a ramp of 1°C/4 s between the annealing and the extension steps, followed by 27 cycles of 94°C for 1 min, 62°C for 1 min, 72°C for 2 min ini-

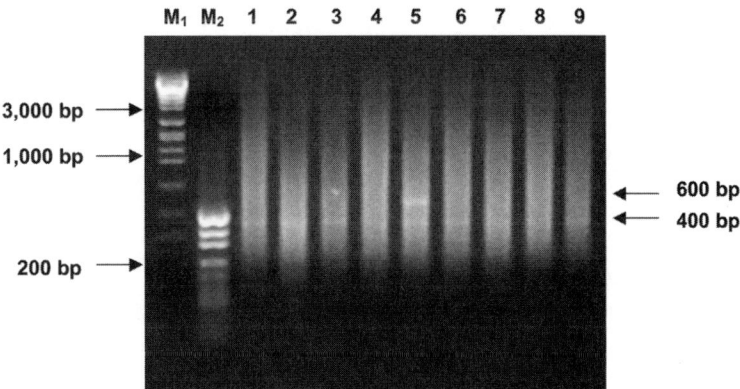

Fig. 2. Electrophoresis of Cy3- or Cy5-labeled single cell DOP-PCR products on 1% agarose gel. Lanes 1–8, Cy3-labeled DOP-PCR products of single lymphocytes from a normal female (*see* **Note 6**); lane 9, Cy5-labeled DOP-PCR products of single male trisomy 18 fibroblast cell; M₁, digested products of SPP-1/*Eco*RI; M₂, digested products of pUC19/*Hpa*II. Note that each labeled product gives a smear ranging in size from 300 bp to 3kb containing a specific band of approx 450 bp and probably another specific band of approx 600 bp.

tially but increased by 14 s for each cycle, and a final extension step at 72°C for 7 min *(16,17)*.

### 3.6. Second-Round of DOP-PCR for Cy3/Cy5 Labeling

1. Transfer 5 µL of the first-round DOP-PCR products (*see* **Note 3**) into a fresh sterilized 0.5-mL PCR tube as DNA template for Cy3/Cy5 labeling (*see* **Note 4**).
2. To the tube, add 5 µL of each of 10X PCR buffer, MgCl₂ (25 m*M*), 0.1% gelatin, and DOP-PCR 6MW primer (20 µ*M*); 4 µL of each of dATP (2 m*M*), dGTP (2 m*M*), and dCTP (2 m*M*); 3 µL of dTTP (2 m*M*); 2 µL of either Cy3-dUTP (1 m*M*) or Cy5-dUTP (1 m*M*); 1 µL of *Taq* polymerase (5 U/µL); and 10 µL of Ultrapure H₂O.
3. Briefly centrifuge the sample; place in a Minicycler (MJ Research); denature at 95°C for 4 min; and cycle for 25 cycles of 94°C for 1 min, 62°C for 1 min, 72°C for 2 min initially but increased by 14 s for each cycle. Add an extension step at 72°C for 10 min at the end.

### 3.7. Purification, Electrophoresis, and Spectrophotometry of Labeled Products

1. Purify the labeled products using an UltraClean PCR Clean-up DNA purification kit (Mo Bio) according to the manufacturer's instructions, and elute with 50 µL of elution buffer (10 m*M* Tris, pH 8.0, DNase free). Electrophorese 5 µL of each of the purified products on a 1% agarose gel to check the quality of both amplification and labeling (*see* **Fig. 2** and **Note 5**).

2. Use 2 μL of each of the purified products for spectrophotometry to obtain both the concentration and purity of the labeled products.

## 3.8. Preparation of Probe Mixtures

1. Prepare the probe mixture in a sterilized 0.5-mL PCR tube containing equal volumes (10 μL, approx 1 μg) of each of Cy3-labeled (test) and Cy5-labeled (reference) DOP-PCR products (*see* **Note 6**), 70 μg of human Cot-1 DNA, 20 μg of sheared salmon sperm DNA, 2 vol of 100% ethanol, and 1/10 vol of 3 *M* NaAC (pH 5.2) (*see* **Note 7**).
2. Place the tube containing the DNA mixture at –20°C for 2 h, and then collect the DNA pellets by centrifuging the tube at 14,000 rpm (10,000*g*) for 25 min at 4°C.
3. Wash the DNA pellets with 70% ethanol, and collect the DNA pellets by centrifuging at 14,000 rpm (10,000*g*) for 10 min at 4°C.
4. Dry the DNA pellets either in an oven at 56°C for 30 min or by air in the dark.
5. Dissolve the dried DNA pellets in 10 μL of hybridization solution.
6. Denature the probes at 80°C for 10 min in a PCR machine, and then preanneal the probes at 37°C for 80 min.

## 3.9. Preparation of Array Slides and Hybridization Chamber

1. Switch on the slide-warming tray and set to 37°C for 30 min prior to preparing the hybridization.
2. Disassemble the hybridization chamber (Corning) and place the base, the cover, and the metal clips on the warm tray.
3. Place the array slide (with the printed array side up) and cover slip on the warm tray.

## 3.10. Preparation for Hybridization

1. Pipet the preannealed probes onto the array area of the array slide sitting on the warm tray and cover with a cover slip.
2. Place the array slide (DNA up) in the base of the hybridization chamber.
3. Pipet 10 μL of hybridization solution into the humidifying well at each end of the chamber base (*see* **Note 8**).
4. Place the cover of the chamber over the base by aligning the base's posts with the cover's indentations.
5. Snap a metal retaining clip onto each side of the chamber while maintaining the chamber on a horizontal plane.
6. Place the hybridization chamber in a cell culture incubator at 37°C for 8–16 h.

## 3.11. Posthybridization Washing

1. Switch on a water bath and set to 45°C.
2. Prepare all the washing solutions and filter them through Acrodisc Syringe Filters (0.2 μm) (Pall).
3. Aliquot solutions into seven Coplin jars (50–60 mL/jar).
4. Leave Coplin jars #1, #6, and #7 at room temperature, and place Coplin jars #2–5 in the water bath once the temperature reaches 45°C.

5. Incubate Coplin jars #2–5 at 45°C for at least 30 min prior to washing.
6. Carefully remove the hybridized slide from the hybridization chamber.
7. Immerse the array slide in Coplin jar #1 (50% formamide/2X SSC) at room temperature in the dark until the cover slip slides off. (Alternatively, use fine-tip forceps to grab the edge of the cover slip and gently lift it up.)
8. Transfer the slide to Coplin jar #2 (50% formamide/2X SSC), and incubate it in this jar at 45°C for 10 min.
9. Transfer the slide to Coplin jar #3 (50% formamide/2X SSC), and incubate it in this jar at 45°C for 10 min.
10. Transfer the slide to Coplin jar #4 (2X SSC), and incubate it in this jar at 45°C for 5 min.
11. Transfer the slide to Coplin jar #5 (2X SSC), and incubate it in this jar at 45°C for 5 min.
12. Transfer the slide to Coplin jar #6 (1X SSC), and place this jar on an orbital mixer (shaking horizontally at 160 rpm) at room temperature for 10 min.
13. Agitate the slide in Coplin jar #7 ($H_2O$) at least 30 times with an up-and-down motion.
14. Dry the slide either using a Microarray high-speed centrifuge (TeleChem) or by air in the dark (*see* **Note 9**). After drying, the slides can be either scanned immediately or stored in a slide box at room temperature in the dark for up to 2 mo (*see* **Note 10**).

## 3.12. Array Scanning

1. Carry out a Preview Scan (40-µ resolution) using a GenePix 4000B scanner (Axon) to locate the array area on the slide (*see* **Note 11**).
2. Perform a Data Scan (10-µ) to acquire the fluoresent images of the hybridized arrays (*see* **Note 11**). These images are the single-wavelength images, and by default they are saved as 16-bit grayscale Tagged Image File Format (TIFF) in a single multi-image, including the ratio (Cy5/Cy3) image saved in both TIFF and Joint Photographic Experts Group (JPEG) format. TIFF files are used for analysis and JPEG files only for presentations.

## 3.13. Data Analysis

The 16-bit grayscale TIFF ratio images are analyzed using GenePix Pro 4.0.1.17 (Axon). Briefly, GenePix Pro 4.0.1.17 uses GenePix Array List (GAL) files to locate the size and position of all spots. The analyzed results are saved as GenePix Results Formats (GPR) files, which contain a header consisting of general information about image acquisition and analysis as well as the data extracted from each feature including more than 40 different parameters. Data analysis can be performed using our custom-made software called GenoData, which imports the GenePix GPR text file and produces data for interpretation of array CGH experiments. Alternatively, data analysis can be carried out manually using a GPR-exported Excel file. This method is described in **Subheadings 3.13.1.–3.13.4.**

### 3.13.1. Exclusion of Dots for Analysis

Seven different parameters in the GPR files are used for data filtering: Dia., %B635 + 2 SD, %B532 + 2 SD, F635 % Sat., F532 % Sat., SNR635, and SNR532. The definitions of these seven parameters are available from http://www.axon.com/gn_GenePix_File_Formats.html (Axon), and they are also given in detail in **Note 12**. Dots are excluded from analysis if they fail to pass any of the following parameters: (1) Dia. >50, (2) %B635 + 2 SD >70, (3) %B532 + 2 SD >70, (4) SNR635 >3.0, (5) SNR532 >3.0, (6) F635 % Sat. = 0, and (7) F532 % Sat. = 0 (*see* **Notes 12** and **13**).

### 3.13.2. Calculation of Normalization Factor, Normalized Ratios, and $log_2$ Ratios

The median of pixel-by-pixel ratios (Cy3/Cy5) of pixel intensities with the median background subtracted is used for interpretation of the results (*see* **Note 14**). The mean of ratios for each chromosome is calculated from up to eight qualified dots. Normalization is then carried out using the 22 means of ratios of all autosomes assuming that the mean ratio value of all autosomes in each hybridization is 1.0 (*see* **Note 15**). This normalization method is available from http://www.axon.com/mr_Axon_KB_Article.cfm?ArticleID=50 (Axon) and can be briefly described as follows:

1. Averagethe median of ratios for all included dots for each chromosome to give the raw mean.
2. Determine the Log value for each raw mean of median of ratios value.
3. Calculate the average of all of the Log values ("Avglog").
4. Calculate the true average ("TrueAvg") (TrueAvg = 10^Avglog).
5. Determine the normalization factor (NF) (NF = 1/TrueAvg).
6. Multiply NF by the raw mean of median of ratios to give the normalized ratios.
7. Calculate the $log_2$-transformed value of the normalized ratios to produce the $log_2$ ratio for every chromosome.

### 3.13.3. Calculation of SD of Normalized Ratios

1. Calculate normalized ratios for every dot by using NF times the median of ratios (normally column AC in a GPR-exported Excel file).
2. Calculate the SD of normalized ratios for each chromosome using only the qualified dots of the relevant chromosomes.

### 3.13.4. Calculation of SD of $log_2$ Ratios

Two methods can be used and they both produce the same results. The first method is as follows:

1. Calculate $log_2$-transformed values of the median of ratios for every dot (e.g., $log_2$-transformed values of column AC in a GPR-exported Excel file).

2. Calculate the SD of $\log_2$-transformed values for each chromosome using only the qualified dots of the relevant chromosomes.

Alternatively, the following method can be used:

1. Calculate the normalized ratios for every dot using NF (e.g., NF times the values of column AC in a GPR-exported Excel file).
2. Calculate the $\log_2$-transformed values of the normalized ratios of all dots.
3. Calculate the SD of $\log_2$-transformed values for each chromosome using only the qualified dots of the relevant chromosomes.

### 3.14. Interpetation of Array CGH Results

Normally, the reference material (*see* **Note 6**) is labeled with Cy5, the test single cell is labeled with Cy3, and the ratio Cy3/Cy5 is reported *(16)*. This type of ratio is employed in most array CGH studies and, hence, it is used in this chapter to describe the diagnostic criteria of the array CGH approach reported in this study; however, for a dye-swap experiment, the ratio of Cy5/Cy3 can be performed in which the test single cell is labeled with Cy5 and the reference material with Cy3. The diagnostic criteria for aneuploidies (e.g., trisomies and monosomies) can be described as follows:

1. Ratios of >1.25 suggest gain of copy number (e.g., trisomies; *see* **Fig. 3B** and **Note 16**) of the relevant chromosomes in the test cell.
2. Ratios of <0.75 suggest loss of copy number (e.g., monosomies, nullisomies) of the relevant chromosomes in the test cell.
3. Ratios of 0.75–1.25 suggest (a) that there is no difference in the copy numbers of the relevant chromosomes between the test and reference samples, indicating that the test sample has the same (diploid) karyotype as the reference; or (b) that the test cell carries a balanced chromosomal abnormality (e.g., polyploidy).

The diagnostic criteria using the $\log_2$ ratios can also be used and are described as follows (*see* **Note 17**):

1. $\log_2$ ratios of >0.32 (equivalent to ratios of >1.25) suggest gain of copy number (e.g., trisomies; *see* **Fig. 3C**) of the relevant chromosomes in the test cell.
2. $\log_2$ ratios of <–0.41 (equivalent to ratios of <0.75) suggest loss of copy number (e.g., monosomies, nullisomies) of the relevant chromosomes in the test cell.
3. $\log_2$ ratios of 0.32 to 0.41 (equivalent to ratios of 0.75–1.25) suggest that in most cases there is no difference in the copy number of the relevant chromosomes between the test and reference samples, indicating that the test sample has the normal diploid karyotype or in rare cases that the test cell carries balanced chromosomal abnormalities (e.g., polyploidy).

## 4. Notes

1. Our custom-made arrays *(16)* contain repeat-depleted human chromosome–specific DNA libraries (SCLs) for all 22 autosomes ($SCL_1$–$SCL_{22}$) and the X chromosome

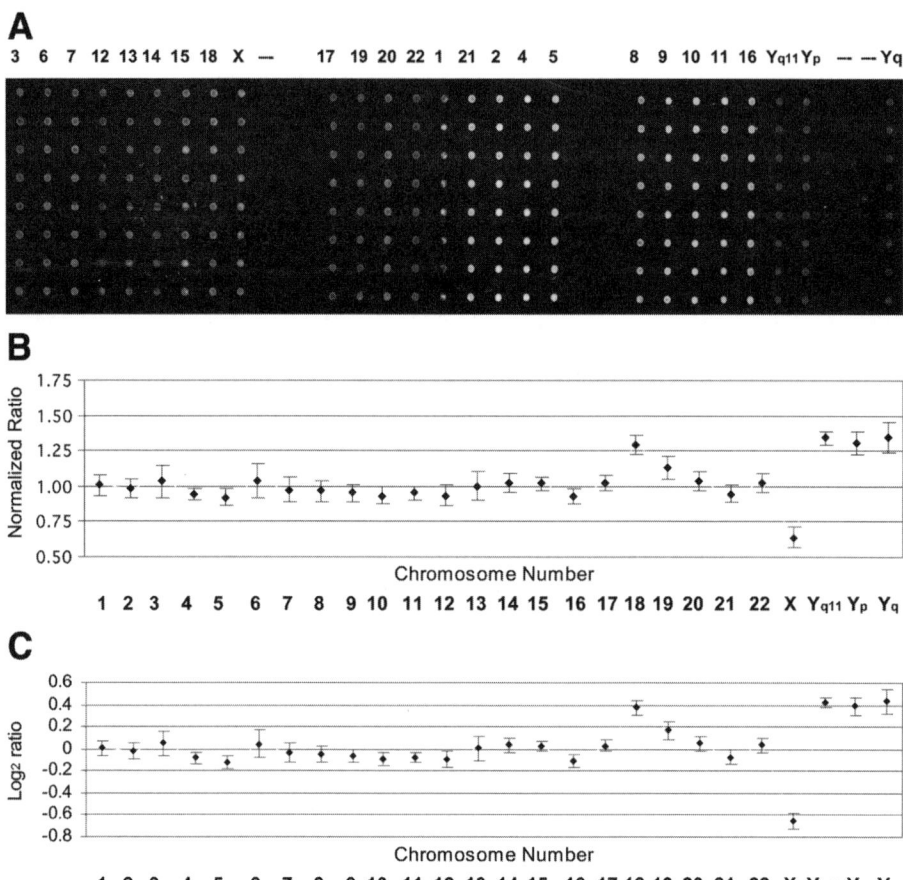

Fig. 3. Diagnosis of single male trisomy 18 (47,XY,+18) fibroblast cell using single cell array CGH analysis. **(A)** JPEG image of Cy3 and Cy5 fluorescence intensities obtained from array CGH experiment of Cy5-labeled DOP-PCR products of a single male trisomy 18 fibroblast cell vs pooled mixture of Cy3-labeled DOP-PCR products from 5–10 normal female single lymphocytes (*see* **Note 6**) captured using GenePix 4000B scanner. The origin of each probe is indicated above each column. Each column has eight identical replicate dots of the same probe. **(B)** Graphic representation of normalized linear ratios (Cy5/Cy3) obtained after analysis of image shown in (A) by GenePix Pro 4.0.1.12. As expected, four probes (SCL$_{18}$, Yq11.2 [called Yq11 in the figure owing to lack of space], Yp, and Yq) give a ratio of >1.25. A ratio of <0.75 for SCLx, and ratios in the range of 0.751.25 of all other autosomal SCLs was observed. These results are consistent with the expected karyotype of the fibroblast cell as 47,XY,+18 (*see* **Note 16**). **(C)** Graphic representation of Log$_2$-transformed ratios (Cy5/Cy3) of normalized linear ratios shown in (B). The expected karyotype of 47,XY,+18 was obtained as shown by Log$_2$-transformed ratios in the range of –0.41–0.32 for all SCLs except the five probes SCL$_{18}$, SCLx, Yq11.2, Yp, and Yq.

(SCL$_X$) *(18)* and three different microdissected Y probes: the short arm of the Y chromosome (Yp), the euchromatic region from the long arm of the Y (Yq11.2), and the heterochromatic region from the long arm of the Y (Yq). DNA microarray slides were spotted (Clive & Vera Ramaciotti Centre for Gene Function Analysis, School of Biotechnology & Biomolecular Sciences, The University of New South Wales, NSW, Australia) with eight replicate spots per probe printed in each column (**Fig. 3A**). Four arrays were printed on each slide (two-by-two configuration) with enough space between them to perform four separate experiments per slide.

2. We have observed aberrant CGH results for single cells that have been frozen at –20°C for many years. This may be owing to residual chemical degradation of the single genome during storage or repeated freeze-thawing under conditions in which constant temperature is not maintained. Best results are obtained with freshly prepared single cells for array CGH analysis.

3. The first-round DOP-PCR product is normally used immediately for Cy3/Cy5-labeling; however, our results indicate that these products can be stored at –20°C for at least 16 mo prior to labeling with no change in the resulting array CGH diagnosis.

4. Repeated rounds of freeze-thawing of the Cy3- and Cy5-dUTPs results in poor labeling. We recommend subaliquoting Cy-dUTPs into single-use portions and handling all solutions containing fluorophores and the hybridized slides in reduced light to minimize photobleaching of Cy3 and Cy5 moieties by ambient light.

5. The labeled second-round DOP-PCR products of single cells give a smear ranging in size from 300 bp to 3 kb on a 1% agarose gel (**Fig. 2**). In addition, after amplification of single lymphocytes a specific band of approx 450 bp and sometimes a 600-bp band can be seen. From single blastomeres (the cells within the early 8- to 12-cell human embryo), two specific bands of approx 600 and 1100 bp may be seen, and these specific bands are reported to be of mitochondrial DNA origin *(2)* and theoretically do not hybridize to the probes on the arrays of genomic DNA origin. The presence of these bands indicates high-quality amplification and labeling. Our experience has shown that Cy3/Cy5-labeled PCR products can be stored at –20°C for at least 2 to 3 mo prior to hybridization with no change in the resulting array CGH diagnosis.

6. The change of reference material from a single cell to a pooled mixture of 5–10 single cell DOP-PCR reactions resulted in a better ratio profile, because in this way individual PCR variations within the single cell reference material can be averaged, resulting in fewer false positive and negative results. We recommend generating the reference material by this method.

7. Although our custom-made arrays use repeat-depleted SCLs *(16)*, human Cot-1 DNA is still required in the hybridization (preannealing) to suppress the repetitive sequences amplified in the test and reference sample (single cells). We generally use 70 µg of human Cot-1 DNA per hybridization, but we have found promising results with 20 µg.

8. It is essential to maintain humidity in the hybridization chamber. We have found that formamide is essential, because the addition of salt solutions (or water) produces poor results but using hybridization buffer (with formamide) works. If a hybridization

chamber is not available, one can use a vented, 138-mm-diameter lightproof opaque Petri dish with 0.5 mL of hybridization solution pipetted around the circumference and placed into a cell culture incubator at 95% humidity. Without the inclusion of hybridization solution surrounding the slide, it is impossible to wash nonspecific contamination off the array area, resulting in an unacceptably high background.

9. Optimal results are obtained using a Microarray High-Speed Centrifuge (TeleChem) or a plate centrifuge to dry slides, because this gives a rapid and streak-free end result. Ambient drying (in the dark) can take anywhere from 30 min to 2 h, depending on humidity levels.

10. In our hands hybridized slides can be stored at room temperature in the dark for up to 73 d with little or no change in the resulting diagnosis *(16)*.

11. The GenePix 4000B scanner scans Cy3 and Cy5 simultaneously. The scanning settings, including photomultiplier (PMT) voltage, scan area, and laser powers, should be optimized. For this study, we always use laser powers of 100% for both channels and an optimal PMT gain within the range of 500–700 for both channels. An extreme PMT gain of <400 results in a large number of dots that fail to pass the filtering parameters of SNR635 and/or SNR532 (not sufficiently different from background) and, therefore, have to be excluded from analysis. An extreme PMT gain of >800 results in saturation of many dots, which means that they appear "white" and fail to pass the filtering parameters of F635 % Sat. = 0 and/or F532 % Sat. = 0 and are also excluded from analysis *(16)*.

12. Seven different parameters of the GPR files generated by GenePix Pro after analyzing the array CGH hybridized images were used in this study for data filtering: (1) Dia. (the diameter in micrometers of the feature-indicator); (2) > %B635 + 2 SD (the percentage of feature pixels with intensities >2 SD above the background pixel intensity, at wavelength #1 [635 nm, for Cy5]); (3) > %B532 + 2 SD (the percentage of feature pixels with intensities >2 SD above the background pixel intensity, at wavelength #2 [523 nm, for Cy3]), (4) SNR635 (the signal-to-noise ratio at wavelength #1 [635 nm, for Cy5], defined by [Mean Foreground 1 – Mean Background 1]/[SD of Background 1]); (5) SNR532 (the signal-to-noise ratio at wavelength #2 [532 nm, for Cy3], defined by [Mean Foreground 1 – Mean Background 1]/[SD of Background 1]); (6) F635 % Sat. (the percentage of feature pixels at wavelength #1 [for Cy5] that are saturated); and (7) F532 % Sat. (the percentage of feature pixels at wavelength #2 [for Cy3] that are saturated).

13. The filtering criteria used in this study are more stringent compared with others *(15)*. The parameters of % >B635 + 2 SD > 70 and % >B532 +2 SD > 70 are the two most stringent ones of the seven filtering parameters employed in this study; however, in most cases the majority of dots excluded from analysis by these two parameters can also be excluded by the other two parameters of SNR532 > 3.0 and SNR635 > 3.0. In rare cases, many dots may be excluded from the analysis by % >B635 + 2 SD > 70 and % >B532 + 2 SD > 70 but could be included for analysis by SNR532 > 3.0 and SNR635 > 3.0. In these cases, one may need to reduce the stringency of >B635 + 2 SD and % >B532 + 2 SD (e.g., >B635 + 2 SD >50 and % >B532 + 2 SD >50) in order to include as many dots as possible in the final analysis. The bottom line is that all dots failing to pass the filtering parameters of

SNR532 > 3.0 and/or SNR635 > 3.0 should be excluded from the final analysis. With good printing of probes, it is very rare that all eight replicate dots of the same SCLs are excluded from analysis owing to failure to pass the filtering criteria.

14. GenePix Pro calculates five different ratios (ratio of medians, ratio of means, median of ratios, mean of ratios, and regression ratio) and displays all of them in the GPR file. We use median of ratios because the median is less affected by extreme values at either end of the ratio distribution, owing to dust contaminants or other artifacts. Array slides with good-quality hybridization should give very similar values for all five ratios.

15. Manual calculation of NF, normalized ratios and their SDs, and $\log_2$ ratios and their SDs takes 2 to 3 h per hybridization. We have produced a custom-made program called GenoData to carry out this process. It computes all five different ratios and graphically displays the ratios and their SDs.

16. The X chromosome produced the expected results in 100% (21/21) of hybridizations, the autosomes in 98.9% (457/462), and the trisomy 18 chromosome in 100% (4/4). The new Yq11.2 probe shows promise, because it gave expected results in 60% (12/20) of the hybridizations and 80% (16/20) when the lower threshold is set to a less stringent value of <0.83. Shown here is the only instance of the other two probes, Yp and Yq, producing the expected result (1/20). This is not surprising because the Yp probe contains the pseudoautosomal region and the Yq probe contains the heterochromatic region of the Yq which are repetitive and nonunique sequences.

17. To be consistent with the linear ratio thresholds of 1.25 and 0.75, Wessendorf et al. *(19)* set the $\log_2$ ratio thresholds for low-copy-number gains and losses to be $\log_2$ 0.32 and $\log_2$ –0.41, respectively. The same diagnostic criteria are employed in our single cell array CGH analysis. It remains to be determined exactly where the cutoff thresholds for our DNA array should lie.

## Acknowledgments

We wish to thank Dr. A. Bolzer from the Institute für Anthropologie und Humangenetik, LMU, München, Germany; and Dr. J. M. Craig from Murdoch Children's Research Institute, Melbourne, Australia for their generous gift of repeat-depleted libraries. We are indebted to the Clive & Vera Ramaciotti Centre for Gene Function Analysis, School of Biotechnology & Biomolecular Sciences, The University of New South Wales, NSW, Australia for their expert DNA-arraying skills. We also thank Dr. A. Connolly from the Microarray Facility, The University of Adelaide, Australia for help with, and access to, the array scanner and software. This work was supported by an Australian Biotechnology Innovation Fund Grant (BIF03028) and BioInnovation SA preseed grant to Reproductive Health Science Pty. Ltd.

## References

1. Wells, D. and Delhanty, J. D. (2000) Comprehensive chromosomal analysis of human preimplantation embryos using whole genome amplification and single cell comparative genomic hybridization. *Mol. Hum. Reprod.* **6,** 1055–1062.

2. Voullaire, L., Slater, H., Williamson, R., and Wilton, L. (2000) Chromosome analysis of blastomeres from human embryos by using comparative genomic hybridization. *Hum. Genet.* **106,** 210–217.

3. Munne, S. (2002) Preimplantation genetic diagnosis of numerical and structural chromosome abnormalities. *Reprod. Biomed. Online* **4,** 183–196.

4. Gianaroli, L., Magli, M. C., Ferraretti, A. P., and Munne, S. (1999) Preimplantation diagnosis for aneuploidies in patients undergoing in vitro fertilization with a poor prognosis: identification of the categories for which it should be proposed. *Fertil. Steril.* **72,** 837–844.

5. Griffin, D. K., Wilton, L. J., Handyside, A. H., Atkinson, G. H., Winston, R. M., and Delhanty, J. D. (1993) Diagnosis of sex in preimplantation embryos by fluorescent in situ hybridisation. *BMJ* **306,** 1382.

6. Kuliev, A., Cieslak, J., Ilkevitch, Y., and Verlinsky, Y. (2003) Chromosomal abnormalities in a series of 6,733 human oocytes in preimplantation diagnosis for age-related aneuploidies. *Reprod. Biomed. Online* **6,** 54–59.

7. Munne, S., Bahce, M., Sandalinas, M., et al. (2004) Differences in chromosome susceptibility to aneuploidy and survival to first trimester. *Reprod. Biomed. Online* **8,** 81–90.

8. Munne, S., Magli, C., Bahce, M., et al. (1998) Preimplantation diagnosis of the aneuploidies most commonly found in spontaneous abortions and live births: XY, 13, 14, 15, 16, 18, 21, 22. *Prenat. Diagn.* **18,** 1459–1466.

9. Wells, D., Sherlock, J. K., Handyside, A. H., and Delhanty, J. D. (1999) Detailed chromosomal and molecular genetic analysis of single cells by whole genome amplification and comparative genomic hybridisation. *Nucleic Acids Res.* **27,** 1214–1218.

10. Voullaire, L., Wilton, L., Slater, H., and Williamson, R. (1999) Detection of aneuploidy in single cells using comparative genomic hybridization. *Prenat. Diagn.* **19,** 846–851.

11. Wilton, L., Williamson, R., McBain, J., Edgar, D., and Voullaire, L. (2001) Birth of a healthy infant after preimplantation confirmation of euploidy by comparative genomic hybridization. *N. Engl. J. Med.* **345,** 1537–1541.

12. Pinkel, D., Segraves, R., Sudar, D., et al. (1998) High resolution analysis of DNA copy number variation using comparative genomic hybridization to microarrays. *Nat. Genet.* **20,** 207–211.

13. Snijders, A. M., Nowak, N., Segraves, R., et al. (2001) Assembly of microarrays for genome-wide measurement of DNA copy number. *Nat. Genet.* **29,** 263, 264.

14. Solinas-Toldo, S., Lampel, S., Stilgenbauer, S., et al. (1997) Matrix-based comparative genomic hybridization: biochips to screen for genomic imbalances. *Genes Chromosomes Cancer* **20,** 399–407.

15. Veltman, J. A., Schoenmakers, E. F., Eussen, B. H., et al. (2002) High-throughput analysis of subtelomeric chromosome rearrangements by use of array-based comparative genomic hybridization. *Am. J. Hum. Genet.* **70,** 1269–1276.

16. Hu, D. G., Webb, G., and Hussey, N. (2004) Aneuploidy detection in single cells using DNA array-based comparative genomic hybridization. *Mol. Hum. Reprod.* **10,** 283–289.

17. Telenius, H., Carter, N. P., Bebb, C. E., Nordenskjold, M., Ponder, B. A., and Tunnacliffe, A. (1992) Degenerate oligonucleotide-primed PCR: general amplification of target DNA by a single degenerate primer. *Genomics* **13,** 718–725.
18. Bolzer, A., Craig, J. M., Cremer, T., and Speicher, M. R. (1999) A complete set of repeat-depleted, PCR-amplifiable, human chromosome-specific painting probes. *Cytogenet. Cell. Genet.* **84,** 233–240.
19. Wessendorf, S., Fritz, B., Wrobel, G., et al. (2002) Automated screening for genomic imbalances using matrix-based comparative genomic hybridization. *Lab. Invest.* **82,** 47–60.

# 13

# Microarray Technology for Mutation Analysis of Low-Template DNA Samples

**Chelsea Salvado and David Cram**

## Summary

Microarrays containing oligonucleotide mutation probes are emerging as useful platforms for the diagnosis of genetic disease. Herein, we describe the development and validation of an in-house microarray suitable for the diagnosis of common cystic fibrosis (CF) mutations in low-template DNA samples such as those taken for preimplantation genetic diagnosis and prenatal diagnosis. The success of the CF microarray was based on the ability to generate sufficient target DNA for hybridization to the array probes using either direct polymerase chain reaction (PCR) amplification or whole-genome amplification followed by PCR. From replicate experiments using target DNA carrying known CF mutations, it was possible to define strict diagnostic parameters for the accurate diagnosis of CF. This protocol serves as a general guide for DNA-testing laboratories to develop other microarray platforms that may eventually replace traditional PCR-based genetic testing in the near future.

**Key Words:** Microarrays; cystic fibrosis; oligonucleotide mutation probes; polymerase chain reaction; whole-genome amplification; preimplantation genetic diagnosis; prenatal diagnosis.

## 1. Introduction

Clinical genetics studies of the last century have identified more than 8000 heritable genetic diseases with distinct clinical symptoms. Even today, the underlying molecular causes for many of these diseases remain unidentified. The availability of the complete sequence of the human genome (*1*) now pro vides a basis to determine the genes and gene alterations responsible for all heritable diseases. Thus, a better understanding of genotype to phenotype will have profound benefits for an increasing proportion of patients at genetic risk, particularly in the areas of genetic counseling, genetic diagnosis, and medical treatment. Consequently, it is likely that existing DNA diagnostic technologies may soon be insufficient to provide testing for all patients. Microarrays are currently being evaluated as a new DNA diagnostic platform to replace existing

From: *Methods in Molecular Medicine: Single Cell Diagnostics: Methods and Protocols*
Edited by: A. Thornhill © Humana Press Inc., Totowa, NJ

diagnostic technologies used in prenatal diagnosis (PND) and preimplantation genetic diagnosis (PGD).

Laboratories offering PND and PGD currently rely on polymerase chain reaction (PCR) methods to amplify a region of the patient's genome for mutation testing. Depending on the type of DNA variant responsible for causing the disease in question, the relevant PCR amplicons are subsequently subjected to different analytical techniques to detect the presence or absence of specific mutations. DNA variants such as deletions, insertions, and trinucleotide repeat and short tandem repeat (STR) sequences involve a loss or gain of the DNA sequence that may be readily diagnosed by allelic sizing methods. For example, the number of repeat units of an allele may be determined by performing PCR across the repeat region followed by sizing of the PCR amplicons on DNA-sequencing gels. Fluorescent PCR (FL-PCR) can also be used as a means to increase the sensitivity of the procedure, particularly for low-template DNA and single cell analyses *(2)*. Conversely, single point mutations captured in PCR amplicons are directly diagnosed at the nucleotide level using a range of different techniques, including restriction fragment length polymorphism analysis, fluorescent single nucleotide primer extension (FL-SNuPE), and direct sequencing *(3)*.

Given the range of diagnostic techniques available, it is not uncommon for individual laboratories to employ different protocols for the diagnosis of the same mutation. Although all of these methods of diagnosis deliver high levels of reliability and accuracy, most are limited to the analysis of one to two mutations at a time, often resulting in an inability to offer a complete diagnosis for diseases caused by a large number of different mutations. For example, because there are more than 1000 different mutations known to cause cystic fibrosis (CF) (refer to the Cystic Fibrosis Genetic Analysis Consortium database (CFGAC), http://www.genet.sickkids.on.ca/cftr), this disease currently poses a significant challenge to the presently employed genotyping strategies *(4,5)*. Currently, most laboratories offer testing for the most prevalent CF mutations, which only cover 80–85% of all CF chromosomes *(6)*. Furthermore, as the genetic basis of more and more diseases is discovered, the capacity of individual diagnostic laboratories to offer testing for all requests will be greatly reduced, primarily owing to the considerable work-up required for the development of new diagnostic protocols for each new indication, which is both time-consuming and costly. Hence, much of the research today focuses on the development of molecular diagnostics with the ability to perform a multitude of tests simultaneously.

Microarray analysis is a newly emerging technology that has a diverse number of applications, including expression profiling *(7,8)*, the analysis of gene dosage *(9,10)*, and the diagnosis of DNA variants (mutation detection) *(11–13)*.

Although the use of microarray technology is largely unexplored in diagnostic fields such as PND and PGD, its use would allow the simultaneous analysis of numerous mutation sites and loci, subsequently broadening the diagnostic capacity of both procedures. Microarray-based methods are strong competitors to the conventional gel-based methods of sequence variation analysis, because their introduction would enable a single procedure to be offered to the majority of patients with minimal or no change to methodology.

The microarray platform used for mutation analysis is usually a miniature platform comprising oligonucleotide probes that have been synthesized *in situ* or robotically spotted (printed) and covalently attached on solid supports (glass, coated glass, silicon, or plastic). The identity of each oligonucleotide is defined by its location *(14)*. As the position of the mutation in the genome dictates the probe sequence and thus G/C content, one of the most difficult problems to overcome when designing a microarray platform is the selection of probes that possess similar melting temperatures ($T_m$) and thus exhibit similar hybridization characteristics *(15)*.

Microarray analysis for the detection of single point mutations and small deletions and insertions involves differential hybridization of fluorescently labeled PCR amplicons (target DNA) to sequence-specific DNA probes representing both the wild-type (WT) and disease alleles. Under optimal hybridization conditions, target DNA forms a stable duplex with its homologous probe, and the signal intensity subsequently emitted by that probe is directly proportional to the amount of bound target DNA. However, probe-target duplexes differing by a single base are unstable and thus easily disrupted on washing, consequently reducing the signal intensity emitted by that particular probe. Hence, relative losses of hybridization to sequence-specific probes are indicative of noncomplementarity between target and probe sequences *(16)*. Alternatively, the test sample may be labeled with one fluorescent dye (i.e., Cy3) and hybridized to the array in the presence of a WT reference DNA sample labeled with the other dye (i.e., Cy5), and allelic discrimination subsequently achieved by quantitating relative losses of hybridization signal to perfect match oligonucleotide probes in the test samples relative to the reference sample *(17)*. To monitor specificity, a "tiled microarray" may be constructed, whereby the probe set for each locus includes four probes that differ by a single base at the interrogation site. Furthermore, a set of eight probes may also be used, consisting of probes complementary to both the sense and antisense sequences of the gene *(18–21)*.

DNA microarray analysis is commonly being used in genetic screening processes to assist in the large-scale identification of single nucleotide changes, deletions and insertions, and other minor sequence variants in genes that underlie genetic and infectious diseases *(22)*. Microarray technology has been used for the detection and screening of mutations present in, e.g., the human immun-odeficiency viurs reverse transcriptase and protease genes *(18)*, the *CFTR* gene *(4)*, and the *β-globin* gene *(23)*. Furthermore, microarrays have been used extensively for the diagnosis of mutations in many of the genes associated with cancer, including the *BRCA1* and *BRCA2* genes *(24–27)*, *RET* mutations in multiple endocrine neoplasia *(28)*, β-catenin mutations in human malignancies *(11)*, and *ATM* mutations in lymphoid neoplasia *(12,29)*, and for the analysis of mutations associated with prostate *(13)* and bladder *(30,31)* cancers. In most studies, high-density oligonucleotide arrays harboring thousands of diagnostic probes were purchased from Affymetrix *(12,29–31)*. However, for smaller projects, others chose to construct their array platforms in-house *(11,28)*.

The procedures given here outline the construction of an in-house microar-ray platform for the diagnosis of two common CF mutations, ΔF508 (CTT dele-tion in exon 10 of the *CFTR* gene) and N1303K (C→G transversional point mutation in exon 21 of the *CFTR* gene), in low-template DNA and single cells. The methods described relate to the final optimized conditions (*see* **Note 1**).

## 2. Materials
### 2.1. Construction of Microarray Platform

1. 21-mer oligonucleotide probes modified with a 5N amine group and C12 linker molecule (*see* **Table 1**); amino silane–coated glass slides, e.g., GAPS II Slides (Corning Microarray Technology, Lindfield, NSW, Australia).
2. Arrayer, e.g., GMS 417 Arrayer (Genetic Microsystems).
3. 3X Standard saline citrate (SSC): 0.15 $M$ NaCl, 15 m$M$ Na$_3$C$_6$H$_5$O$_7$·2H$_2$O.
4. Succinic anhydride.
5. 1-Methyl-2-pyrrolidinone.
6. Sodium borate.
7. ddH$_2$O.
8. 95% Ethanol.
9. Glass slide holder.
10. Glass beakers.
11. Falcon tubes (50 mL).
12. Water bath.
13. Heating block.
14. Ultraviolet (UV) crosslinker.
15. Hybridization oven.
16. Orbital shaker.
17. Centrifuge.

### 2.2. Generation of Target DNA

1. Lysis buffer, neutralizing buffer, 10X K$^+$-free PCR buffer, PCR reaction mix (*see* **Table 2**).

**Table 1**
**Oligonucleotide Microarray Probes**

| Name | Sequence (5' - 3') |
|------|---------------------|
| ΔF508 wt | AAATATCATCTTTGGTGTTTC |
| ΔF508 mut | GAAAATATCATTGGTGTTTCC |
| N1303K wt | TTAGAAAAAACTTGGATCCCT |
| N1303K mut | TTAGAAAAAAGTTGGATCCCT |
| N1303K(A) | TTAGAAAAAAATTGGATCCCT |
| N1303K(T) | TTAGAAAAAATTTGGATCCCT |
| GFP | TCGAATCGAACTAAAAGGCAT |

**Table 2**
**Buffers and Reaction Mixtures for Generation of Target DNA**

| Buffer/mix | Components[a] |
|------------|--------------|
| Lysis buffer | 200 m$M$ KOH, 50 m$M$ DTT |
| Neutralizing buffer | 900 m$M$ Tris-HCl (pH 8.3), 300 m$M$ KCl, 200 m$M$ HCl |
| 10X K$^+$-free PCR buffer | 100 m$M$ Tris-HCl (pH 8.3) |
| PCR reaction mix | 1X K$^+$-free PCR buffer, 166 µ$M$ dNTPs, 2.5 m$M$ MgCl$_2$, 0.5 µ$M$ forward primer (F), 0.5 µ$M$ reverse primer (R), 1 U *Taq* DNA Polymerase; final volume 30 µL |

[a]DTT, dithiothreitol.

**Table 3**
**PCR Primers**

| Name | Sequence (5' – 3') |
|------|---------------------|
| ΔF508-1F | GACTTCACTTCTAATGATGAT |
| ΔF508-1R | CTCTTCTAGTTGGCATGC |
| ΔF508-2F | TGGGAGAACTGGAGCCTT |
| ΔF508-2R | GCTTTGATGACGCTTCTGTAT |
| N1303K-1F | AAATGTTCACAAGGGACTCC |
| N1303K-1R | TGATGTCAGCTATATCAGCC |
| N1303K-2F | TCTTTTTTGCTATAGAAAG |
| N1303K-2R | TCTGCAACTTTCCATATTTC |
| PEP | NNNNNNNNNNNNNNNA/C/G/T |

2. PCR primers, primer extension preamplification (PEP)-PCR random primer (*see* **Table 3**).
3. Agarose gel equipment, agarose gel DNA purification kit (i.e., Perfectprep Gel Cleanup Kit; Eppendorf, North Ryde, NSW, Australia).

**Table 4**
**Microarray Hybridization and Washing Buffers**

| Buffer | Components[a] |
|--------|---------------|
| Prehybridization | 5X SSC, 0.1% SDS, 0.1 mg/mL BSA |
| Hybridization | 5X SSC, 0.1% SDS, 0.1 mg/mL salmon sperm DNA |
| Wash 1 | 1X SSC/0.1% SDS |
| Wash 2 | 0.5X SSC |
| Wash 3 | 0.1X SSC |

[a]SDS, sodium dodecyl sulfate; BSA, bovine serum albumin.

## 2.3. Fluorescent Labeling of Target DNA

1. BioPrime DNA Labelling System (Invitrogen Australia, Melbourne, Australia).
2. Klenow fragment.
3. FluoroLink Cy3-dCTP or Cy5-dCTP (Amersham Biosciences, Castle Hill, NSW, Australia).
4. dNTP mix: 1.2 m$M$ dATP, dGTP, and dTTP; 0.6 m$M$ dCTP.
5. 10 m$M$ Tris-HCl (pH 8.0).
6. 1 m$M$ EDTA.
7. TE (pH 7.4 and 8.0).
8. 0.5 $M$ EDTA (pH 8.0).
9. Microcon YM-30 purification columns (Millipore, North Ryde, NSW, Australia).
10. Heating block.
11. Microcentrifuge.

## 2.4. Hybridization and Washing

1. Hybridization and wash buffers (*see* **Table 4**).
2. ddH$_2$O.
3. 100% Ethanol.
4. Coplin jar.
5. Hybridization chamber (Corning Microarray Technology: Crown Scientific, Victoria, Australia).
6. Falcon tubes (50 mL).
7. Cover slips (22 × 22 mm).
8. Water bath.
9. Heating block.
10. Centrifuge.

## 2.5. Microarray Analysis

1. Confocal microscope laser and relevant software, e.g., GMS 418 Array Scanner (Genetic MicroSystems).
2. Analytical software, e.g., ImaGene Software (BioDiscovery).

## 3. Methods

Manual construction of the microarray chip and subsequent microarray analysis require the combination of at least five different components:

1. Chemically modified slides for the attachment of oligonucleotide probes.
2. An arrayer for probe deposition.
3. A system to facilitate hybridization of target DNA.
4. A confocal microscope laser (microarray scanner) to extract the data.
5. A sophisticated software program to interpret the results.

All of these components are now commercially available (*32*). Examples are provided in **Subheading 2.**

### 3.1. Construction of Microarray Platform

#### 3.1.1. Microarray Probes for Mutation Detection

Two oligonucleotide allelic probes were incorporated onto the array for both ΔF508 and N1303K, one of which was complementary to the WT allele and one complementary to the mutant allele. To monitor the specificity of hybridization, two additional probes were also included for N1303K (N1303K[A] and N1303K[T]), each of which differed by an individual base at the interrogation site. A probe complementary to a region of the jellyfish green fluorescence protein (GFP) gene that shares no sequence homology with the human genome was also included as a negative control. All oligonucleotide probes were designed to contain 21 nucleotides with a central interrogation site and modified to include a 5' amine group with a C12 linker molecule. Refer to **Table 1** for all oligonucleotide probe sequences. Prior to probe deposition, oligonucleotides were diluted to 25 μ*M* in 3X SSC buffer.

#### 3.1.2. Probe Deposition

Probes were deposited onto amino silane–coated glass slides using a contact arrayer (ring and pin system). Probes were printed in triplicate to create an allelic probe set, with a distance of 520 μm between each spot. A total of 100 pg of oligonucleotide was deposited onto the slide for each probe spot (*see* **Note 2**). In addition to the GFP probe, PCR-grade H$_2$O, SSC buffer, and blanks were employed as negative controls (**Fig. 1**).

To immobilize the DNA probes (specific for amino silane–coated slides), do the following:

1. Rehydrate the slides by suspending the individual slides upside down in the steam of a boiling water bath approx 20 cm above water level for 10 s.
2. Snap dry on a 100°C hot plate for 3 s.
3. UV crosslink for a total of 300 mJ.

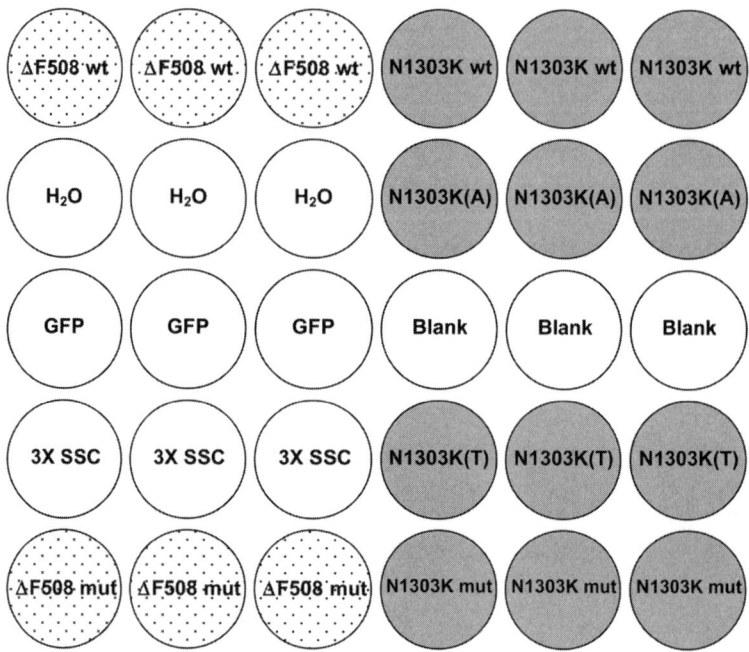

Fig. 1. Schematic representation of microarray platform. The allelic probes of ΔF508 (dotted circles) and N1303K (gray circles) were printed in triplicate onto the array platform, with a distance of 520 μm between each spot. The GFP DNA probe and other controls (white circles) were similarly printed in triplicate at various locations on the slide.

4. Bake at 80°C for 2 h in a hybridization oven.

Reactive silane groups remaining on the slides are blocked to prevent nonspecific binding as follows:

1. Place a maximum of 10 slides in a glass slide holder.
2. In a glass beaker dissolve 5.5 g of succinic anhydride in 350 mL of 1-methyl-2-pyrrolidinone.
3. Immediately add 15 mL of 1 $M$ sodium borate and rapidly dunk the slides five times.
4. Place the beaker containing the slides on an orbital shaker for 15 min.
5. Gently plunge the slides in 95°C ddH$_2$O for 2 min and rinse five times in 95% ethanol.
6. Dry by centrifuging at 45$g$ for 30 s (the slides may be placed in individual Falcon tubes and then dried by centrifugation).
7. Place the printed arrays back in a storage container and store in desiccator at ambient temperature. According to the directions of the slide manufacturer (Corning Microarray Technology), slides may be stored for up to 6 mo prior to hybridization.

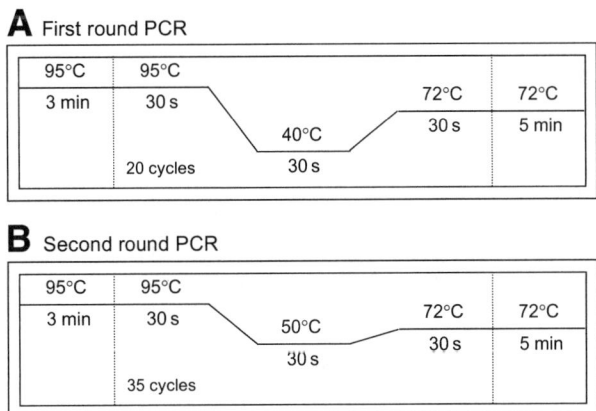

Fig. 2. Thermal cycling conditions for single cell PCR amplification of ΔF508 and N1303K using nested primers. (**A**) First-round amplification using primers F508-1F/R or N1303-1F/R. (**B**) A 2-µL aliquot of the first-round amplification was used as a template in a second-round amplification using primers F508-2F/R or N1303-2F/R, respectively.

### 3.2. Generation of Target DNA

Prior to PCR amplification, 10 and 1 cell samples were lysed in 2.5 µL of lysis buffer at 65°C for 10 min, followed by the addition of 2.5 µL of neutralizing buffer.

### 3.2.1. Direct Amplification from Template

Single cell PCR was employed for the amplification of a single locus for microarray analysis. For example, a region of exon 10 of the *CFTR* gene harboring the ΔF508 mutation was amplified using nested primers (F508-1F/R and F508-2F/R) (**Table 1**) according to the thermal cycling conditions displayed in **Fig. 2**.

### 3.2.2. Indirect Amplification from Template

Direct PCR amplification of target DNA is sufficient for microarray diagnosis of a limited number of mutations. However, in more complex cases requiring the analysis of multiple mutations, single cell multiplex PCR often becomes problematic, primarily owing to nonspecific interactions between an increasing number of primers (*33–36*). Alternatively, whole-genome amplification (WGA), which aims to amplify the entire genome of low-template DNA in an unbiased fashion (*37*), could be used to effectively create a universal target DNA sample covering multiple loci. At present, the most common methods of WGA are PEP-PCR (*38–40*), degenerate oligonucleotide-primed (DOP) PCR (*41*), and multiple displacement amplification (MDA) (*42*). However, our experience has shown that each of these methods exhibits varying degrees of

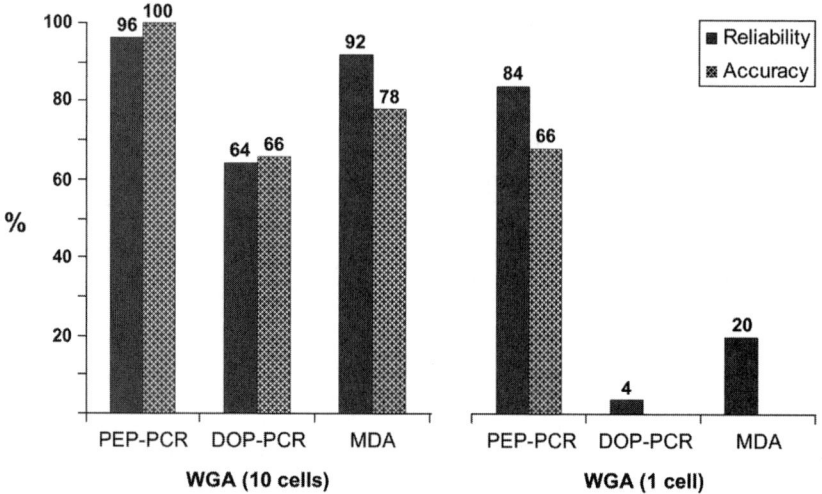

Fig. 3. Reliability and accuracy of WGA on 10-cell and single cell templates. WGA reactions to assess reliability ($n = 5$) were performed according to published protocols and the products subjected to locus-specific PCR amplification of five different *CFTR* exons (4, 9, 10, 11, and 21). Reliability was calculated as the percentage of PCR amplicons present when resolved on 1.5% agarose gels. WGA reactions to assess accuracy were performed on samples heterozygous for either ΔF508 or N1303K (10-cell template, $n = 20$; 1-cell template, $n = 50$). Products were subjected to locus-specific PCR amplification of either exon 10 or 21 of the *CFTR* gene and analyzed using allelic sizing methods (ΔF508) and FL-SNuPE (N1303K). Accuracy was determined by scoring the presence or absence of the two ΔF508 and N1303K alleles. Accuracy was not determined for DOP-PCR and MDA, owing to an inability to reliably generate PCR products from reaction products.

reliability and accuracy, as judged on buccal cells heterozygous for ΔF508 and lymphoblasts heterozygous for N1303K. High reliability and accuracy is mandatory in the clinical setting of PND and PGD, to prevent the occurrence of a misdiagnosis. Thus, for the purposes of this study, a reliability and accuracy level 85% was arbitrarily assigned as a cutoff value for a particular form of WGA to be considered for use in these areas of diagnosis.

As shown in **Fig. 3**, PEP-PCR was the only method of WGA found to provide reliable amplification of single cells (84%), such that DOP-PCR and MDA yielded extremely low reliability rates of 4 and 20%, respectively (consequently, the accuracy of both of these methods was unable to be calculated). However, because PEP-PCR produced accurate results in only 68% of single cell amplifications, it was concluded that WGA at the single cell level did not provide high enough levels of reliability and accuracy to be used in conjunction with microarray analysis. Conversely, PEP-PCR and MDA were both found to exhibit high levels of reliability (≤85%) following the amplification of 10 cell samples. Whereas PEP-PCR displayed a similarly high level of accuracy, MDA was affected by a high rate of allelic dropout (ADO) and thus low accuracy. By contrast, DOP-PCR provided low levels of both reliability and accuracy, deeming this method the most unsuitable for the amplification of low-template DNA. Thus, owing to high reliability (98%) and accuracy (100%), PEP-PCR amplification of 10 cells was the only method of WGA found suitable for the generation of target DNA for microarray analysis in PND and PGD and, consequently, the only method of WGA employed for the generation of target DNA within our study. PEP-PCR was performed according to published protocols *(38–40)*. To generate target DNA, independent locus-specific PCR reactions were subsequently performed using nested primers F508-1F/R and F508-2F/R for ΔF508 and N1303-1F/R and N1303K-2F/R for N1303K (**Table 1**). Refer to **Fig. 2** for the thermal cycling conditions (*see* also **Notes 3** and **4**).

### 3.2.3. Purification of Target DNA

PCR amplicons were generated directly or indirectly via WGA and subsequently resolved on 1.5% ethidium bromide agarose gels. DNA bands were excised under UV light and purified using an agarose gel DNA purification kit according to the manufacturer's directions.

## 3.3. Fluorescent Labeling of Target DNA

Approximately 2 ng of target DNA (PCR amplicons) was fluorescently labeled by random priming using the BioPrime DNA Labelling System.

1. Bring the DNA sample to a final volume of 10.5 µL with TE (pH 8.0).
2. Add 10 µL of 2.5X Random Primer Solution, boil for 5 min, and place on ice.
3. Add 2.5 µL of dNTP mix, 1.5 m$M$ FluoroLink Cy3-dCTP or Cy5-dCTP, and 20 U of Klenow fragment.
4. Incubate labeling reactions at 37°C for 2 h.
5. Terminate the reactions with the addition of 2.5 µL 0.5 $M$ EDTA (pH 8.0).
6. Dilute labeled DNA samples in 400 µL of TE (pH 7.4), and purify and concentrate through a Microcon YM-30 purification column to a volume of 10 µL.

## 3.4. Hybridization and Washing

### 3.4.1. Prehybridization

Prior to hybridization, slides were incubated in 50 mL of prehybridization buffer (**Table 4**) at 37°C for 45 min, rinsed in ddH$_2$O, and dried by centrifugation.

### 3.4.2. Hybridization

1. Add 10 µL of purified labeled target DNA to 15 µL of hybridization buffer (**Table 4**), to make a total volume of 25 µL.
2. Denature at 95°C for 5 min and cool to room temperature.
3. Apply the hybridization mix to the array grid and cover with a 22 × 22 mm cover slip.
4. Place the microarray slide in a hybridization chamber and perform hybridization in a 37°C water bath for 4–16 h.

### 3.4.3. Posthybridization Washes

1. Remove the cover slips.
2. Wash the slide once in wash buffer 1 (Table 4) for 10 min at 37°C.
3. Wash once in wash buffer 2 (Table 4) for 10 min at room temperature.
4. Wash once in wash buffer 3 (Table 4) for 10 min at room temperature.
5. Rinse the slide in 100% ethanol and dry by centrifugation (per the slide manufacturer's recommendation).

## 3.5. Microarray Analysis

### 3.5.1. Scanning and Image Analysis

To obtain an image of the completed hybridization reaction, the microarray slide was immediately scanned with a confocal microscope laser at 543 nm (Cy3) or 633 nm (Cy5). The image data were saved as a tagged image file and subsequently imported into ImaGene 4.0 (BioDiscovery), software capable of extracting the relevant pixel values emitted by target DNA bound to the probes. Prior to analysis, grids were manually placed on top of the image to specify both the location and local background region of each probe. A gene list was also created and linked to the grid, subsequently allowing each fluorescent spot on the array to be identified as a specific probe-target duplex. ImaGene software was used to quantify the signal and local background mean intensity for each probe spot. The true signal intensity (TSI) of each spot was calculated by subtracting the background intensity from the signal intensity. The three TSI values from each allelic probe set were subsequently averaged to give the final allelic signal intensity (FASI). Genotypes were subsequently assigned according to the relative allelic signal intensity (RASI), which is equal to the ratio of the WT FASI to the sum of the WT and mutant FASI values: WT FASI/(WT FASI + mutant FASI). Thus, homozygous WT, homozygous mutant, and heterozygous samples were expected to generate RASI values approaching 1, 0, and 0.5, respectively (**Fig. 4**).

### 3.5.2. Development of Diagnostic Parameters

To develop diagnostic parameters that could be used for the genetic diagnosis of low-template target DNA, five independent hybridization experiments were

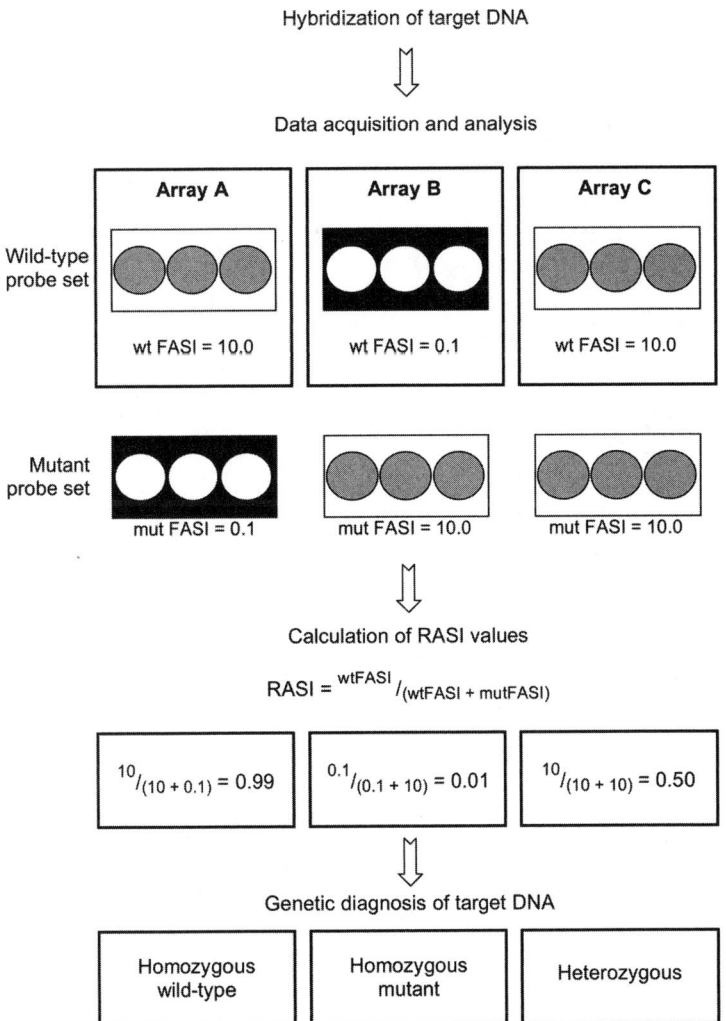

Fig. 4. Calculation of RASI values following hybridization of different target DNAs: homozygous WT (Array A), homozygous mutant (Array B), and heterozygous (Array C). WT (wt) and mutant (mut) allelic probes were printed in triplicate. Gray and white circles represent matched and mismatched probe-target duplexes, respectively. The TSI of each spot (*n* = 3) was calculated, and these values were averaged to give the wt and mut FASI. RASI was subsequently calculated and a genetic diagnosis made, such that homozygous WT, homozygous mutant, and heterozygous samples were expected to generate RASI values approaching 1, 0, and 0.5, respectively. FASI values were arbitrarily assigned for simplicity.

performed for each genotype of ΔF508 and N1303K. Target DNA samples were generated by PCR amplification of genomic DNA with a homozygous WT, homozygous mutant, and heterozygous genotype. PCR amplicons were subsequently fluorescently labeled and hybridized to the microarray. RASI values were calculated and then used to derive diagnostic parameters (based on an acceptable range of RASI values) for homozygous WT, homozygous affected, and heterozygous genotypes (**Table 5**). It was then possible to use this criterion to genotype unknown target DNA samples according to the generated RASI values.

### 3.6. Application of In-house CFTR Microarray Analysis

### 3.6.1. Microarray Diagnosis of ΔF508 Following Direct Amplification

**Figure 5** illustrates the accurate diagnosis of ΔF508 genotypes in PCR amplicons generated from 10-cell (**Fig. 5A**) and single cell (**Fig. 5B**) samples by direct PCR amplification of template DNA. Cell samples were subjected to nested locus-specific PCR amplification of exon 10 of the *CFTR* gene to generate target DNA for microarray analysis. RASI values were calculated and correct diagnoses made according to the diagnostic parameters specified in **Table 5**. Furthermore, clinical samples from a previous PGD case involving ΔF508 were reevaluated using microarray analysis, and these results were compared with those obtained using the original methods of FL-PCR and allelic sizing. Target DNA for microarray analysis was generated by reamplifying first-round PCR products using unlabeled primers and hybridized to the array. Diagnoses for three PGD embryos (E5, E13, and E15) were made according to the diagnostic parameters specified in **Table 5**, achieving 100% concordance with the original results of the clinical case. In addition, microarray-based analysis was sensitive enough to detect the occurrence of preferential amplification in a heterozygous sample (E15), which was similarly detected using allelic sizing (**Fig. 5C**) *(43)*.

### 3.6.2. Microarray Diagnosis of ΔF508 and N1303K Following PEP-PCR

Microarray analysis also provided accurate diagnosis of ΔF508 and N1303K genotypes in 10-cell samples initially amplified by PEP-PCR. Low-template DNA samples were amplified by PEP-PCR, and the products were subjected to two independent locus-specific PCR reactions for the amplification of exons 10 and 21 of the *CFTR* gene. PCR amplicons were combined, fluorescently labeled, and hybridized to the microarray. RASI values were calculated and diagnoses made according to the diagnostic parameters specified in **Table 5**. **Figure 6A** shows the microarray results of low-template DNA heterozygous for ΔF508 (N1303K homozygous WT), and **Fig. 6B** shows the results of low-template DNA heterozygous for N1303K (ΔF508 homozygous WT) (*see* **Note 5**).

For all target DNA samples analyzed, correct ΔF508 and N1303K genotypes were assigned using the in-house microarray system.

**Table 5**
**Definition of Diagnostic Parameters for ΔF508 and N1303K Genotypes for Oligonucleotide Microarray Analysis**

| Mutation site | Target DNA genotype | RASI | | | | | Mean RASI | Acceptable RASI[a] |
|---|---|---|---|---|---|---|---|---|
| | | 1 | 2 | 3 | 4 | 5 | | |
| ΔF508 | Homozygous WT | 0.912 | 0.867 | 0.820 | 0.817 | 0.818 | 0.847 | 0.763–1.000 |
| | Homozygous mutant | 0.020 | 0.010 | 0.026 | 0.029 | 0.029 | 0.023 | 0.000–0.039 |
| | Heterozygous | 0.524 | 0.621 | 0.558 | 0.599 | 0.477 | 0.556 | 0.440–0.671[b] |
| N1303K | Homozygous WT | 0.842 | 0.875 | 0.889 | 0.871 | 0.722 | 0.840 | 0.704–1.000 |
| | Homozygous mutant | 0.080 | 0.108 | 0.072 | 0.078 | 0.075 | 0.083 | 0.000–0.112 |
| | Heterozygous | 0.467 | 0.420 | 0.483 | 0.464 | 0.526 | 0.472 | 0.396–0.584[b] |

[a]Mean ± 2 SE (adjusted for diagnostic purposes).
[b]RASI value falling outside the acceptable range indicates a heterozygous sample associated with preferential amplification of the WT (approaching 1) or mutant (approaching 0) allele.

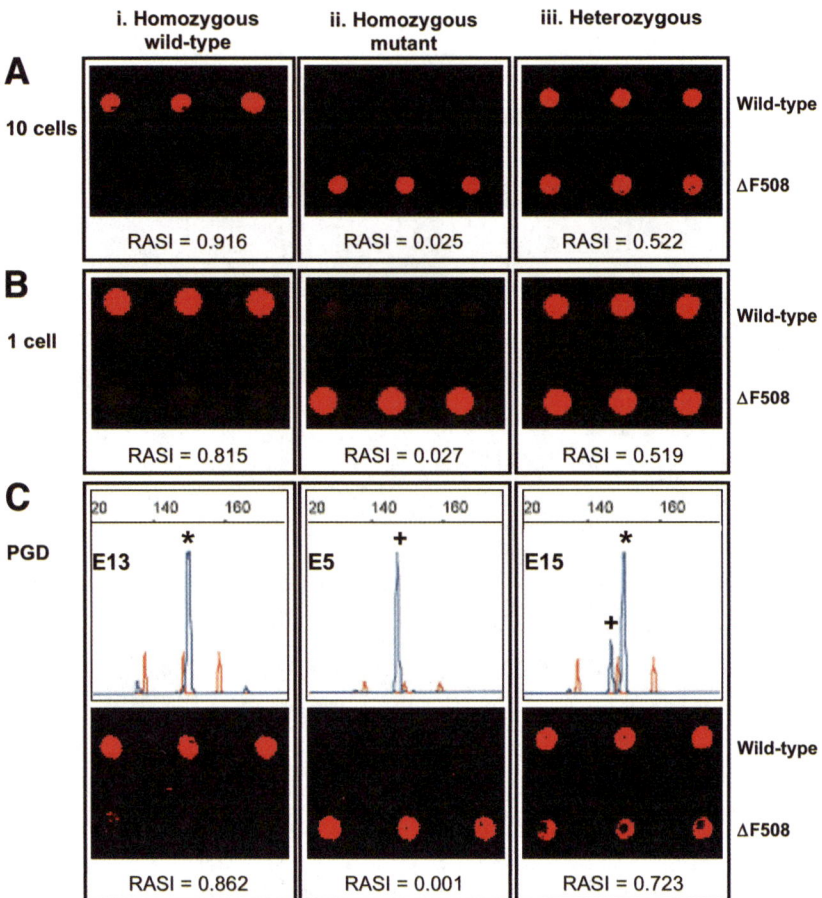

Fig. 5. Microarray detection of ΔF508 following direct PCR amplification of template DNA. Samples containing (**A**) 10 and (**B**) single buccal cells were subjected to locus-specific PCR amplification of exon 10 of the CFTR gene to generate target DNA. (**C**) Comparative diagnosis of ΔF508 in embryos using allelic sizing and microarray methods. (*Top*) First- and second-round FL-PCR and allelic sizing methods were used in PGD to genotype embryos. Allelic peaks are marked as follows: + (ΔF508 allele) and * (WT allele). Unmarked peaks represent internal molecular weight markers (*left to right*): 139, 150, and 160 bp. (*Bottom*) PGD samples were reevaluated using microarray analysis. Target DNA was generated by reamplifying first-round PCR products (*43*).

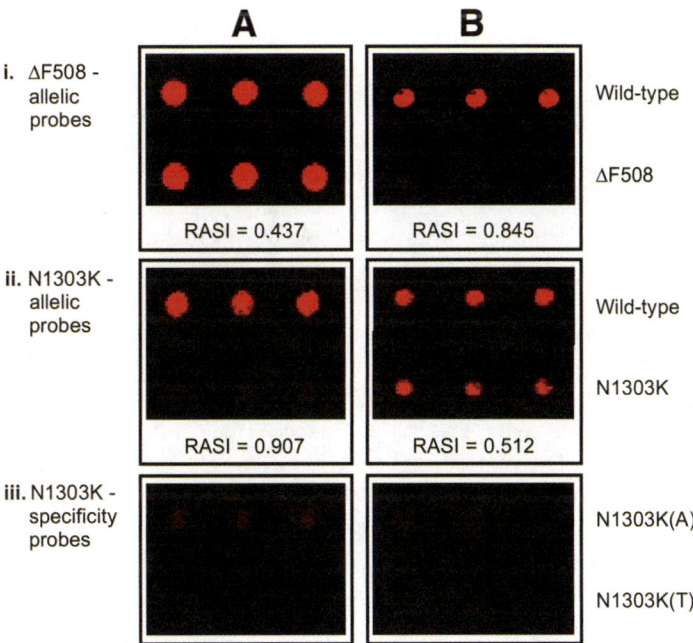

Fig. 6. Microarray detection of ΔF508 and N1303K in low-template DNA amplified by WGA. Samples containing (**A**) 10 buccal cells heterozygous for ΔF508 and homozygous WT for N1303K and (**B**) 10 lymphoblasts homozygous WT for ΔF508 and heterozygous for N1303K were amplified by PEP-PCR, and the products were subjected to locus-specific PCR amplification of exons 10 and 21 of the CFTR gene to generate microarray target DNA.

## 4. Notes

1. Most of the commonly diagnosed CF mutations (apart from ΔF508) are single point mutations (CFGAC database). Accordingly, the microarray protocol described should be applicable to the detection of most other clinically significant CF mutations. For example, we have also shown that G551D (G → A, transitional point mutation) can be diagnosed using our optimized array system. However, one set of hybridization and wash conditions may not provide optimal target discrimination for all probes on the microarray (**44**). In addition, because it is not currently possible to accurately predict the performance of each probe by its primary DNA sequence, the need for experimental screening of probes cannot be eliminated (**45**). Thus, further optimization may be required with respect to probe length, position of the interrogation site within the probe, choice of microarray platform (slide type), and hybridization and wash conditions.

2. For larger projects, it is advisable to print the same probe at two or more different locations on the chip to increase the sensitivity and specificity of the results, by

compensating for nonspecific signals that may arise from localized surface imperfections *(29)*.

3. Direct use of WGA products as target DNA for microarray analysis (i.e., without locus-specific PCR) resulted in high levels of nonspecific hybridization and background fluorescence. Hence, indirect locus-specific PCR amplification of WGA products was employed to generate PCR amplicons for microarray analysis. WGA techniques yield high amounts of DNA, thus enabling the products of such reactions to be used as template DNA in multiple downstream PCR reactions as well as the analysis of many independent loci without having first to develop problematic multiplex PCR conditions.

4. Using the entire human genome as target DNA for microarray analysis is currently hampered by two problems: first, ubiquitously distributed repetitive DNA sequence elements tend to hybridize to multiple loci, significantly increasing the chance of nonspecific hybridization; and, second, the frequency of a given single-copy sequence and therefore the likelihood of meeting its complementary probe is low, even when using a high concentration of target DNA *(46)*. Thus, there have been very few reports demonstrating the use of the entire genome as microarray target DNA for mutation detection. However, a recent publication describes the use of linker adapter PCR to simplify total genomic DNA for use as target DNA prior to microarray analysis of single-nucleotide polymorphisms (SNP) *(47)*.

5. Microarray technology could also be used for the analysis of ΔF508 and N1303K in single cells amplified by WGA, achieving concordant results with those obtained using allelic sizing methods and FL-SNuPE. However, both methods detected a high rate of ADO, as described in **Fig. 3**. Thus, as with any diagnostic method of PND and PGD, the presence of ADO must be monitored to prevent the occurrence of a misdiagnosis. Linked polymorphic markers, such as STR sequences, are most often used in single cell PCR to identify ADO and extraneous DNA contamination *(48–50)*. However, the sensitivity of microarray analysis is lessoned in targets comprising short repeats, owing to increased nonspecific hybridization of the target DNA to the diagnostic probes *(26)*. Therefore, the most effective way to monitor ADO in a microarray assay would be to incorporate probes for the detection of SNPs linked to disease-specific mutations *(51)*. Probes for SNP detection would be designed and thus should behave similarly to the ΔF508 and N1303K probes used for allelic discrimination.

## References

1. Lander, E. S., Linton, L. M., Birren, B., et al. (2001) Initial sequencing and analysis of the human genome. *Nature* **409,** 860–921.
2. Wells, D. and Sherlock, J. K. (1998) Strategies for preimplantation genetic diagnosis of single gene disorders by DNA amplification. *Prenat. Diagn.* **18,** 1389–1401.
3. Handyside, A. H. and Delhanty, J. D. (1997) Preimplantation genetic diagnosis: strategies and surprises. *Trends Genet.* **13,** 270–275.
4. Cronin, M. T., Fucini, R. V., Kim, S. M., Masino, R. S., Wespi, R. M., and Miyada, C. G. (1996) Cystic fibrosis mutation detection by hybridization to light-generated DNA probe arrays. *Hum. Mutat.* **7,** 244–255.

5. Southern, E. M. (1996) DNA chips: analysing sequence by hybridization to oligonucleotides on a large scale. *Trends Genet.* **12,** 110–115.

6. Gilbert, F. (2001) Cystic fibrosis carrier screening: steps in the development of a mutation panel. *Genet. Test* **5,** 223–227.

7. Kawasaki, E. S. (2004) Microarrays and the gene expression profile of a single cell. *Ann. NY Acad. Sci.* **1020,** 92–100.

8. Ginsberg, S. D., Elarova, I., Ruben, M., et al. (2004) Single-cell gene expression analysis: implications for neurodegenerative and neuropsychiatric disorders. *Neurochem. Res.* **29,** 1053–1064.

9. Fiegler, H., Carr, P., Douglas, E. J., et al. (2003) DNA microarrays for comparative genomic hybridization based on DOP-PCR amplification of BAC and PAC clones. *Genes Chromosomes Cancer* **36,** 361–374.

10. Hu, D. G., Webb, G., and Hussey, N. (2004) Aneuploidy detection in single cells using DNA array-based comparative genomic hybridization. *Mol. Hum. Reprod.* **10,** 283–289.

11. Kim, I. J., Kang, H. C., Park, J. H., et al. (2003) Development and applications of a beta-catenin oligonucleotide microarray: beta-catenin mutations are dominantly found in the proximal colon cancers with microsatellite instability. *Clin. Cancer Res.* **9,** 2920–2925.

12. Fang, N. Y., Greiner, T. C., Weisenburger, D. D., et al. (2003) Oligonucleotide microarrays demonstrate the highest frequency of ATM mutations in the mantle cell subtype of lymphoma. *Proc. Natl. Acad. Sci. USA* **100,** 5372–5377.

13. Dumur, C. I., Dechsukhum, C., Ware, J. L., et al. (2003) Genome-wide detection of LOH in prostate cancer using human SNP microarray technology. *Genomics* **81,** 260–269.

14. Tillib, S. V. and Mirzabekov, A. D. (2001) Advances in the analysis of DNA sequence variations using oligonucleotide microchip technology. *Curr. Opin. Biotechnol.* **12,** 53–58.

15. Bains, W. (1994) Selection of oligonucleotide probes and experimental conditions for multiplex hybridization experiments. *Genet. Anal. Tech. Appl.* **11,** 49–62.

16. Bednar, M. (2000) DNA microarray technology and application. *Med. Sci. Monit.* **6,** 796–800.

17. Hacia, J. G. and Collins, F. S. (1999) Mutational analysis using oligonucleotide microarrays. *J. Med. Genet.* **36,** 730–736.

18. Kozal, M. J., Shah, N., Shen, N., et al. (1996) Extensive polymorphisms observed in HIV-1 clade B protease gene using high-density oligonucleotide arrays. *Nat. Med.* **2,** 753–759.

19. Sapolsky, R. J., Hsie, L., Berno, A., Ghandour, G., Mittmann, M., and Fan, J. B. (1999) High-throughput polymorphism screening and genotyping with high-density oligonucleotide arrays. *Genet. Anal.* **14,** 187–192.

20. Lipshutz, R. J., Fodor, S. P., Gingeras, T. R., and Lockhart, D. J. (1999) High density synthetic oligonucleotide arrays. *Nat. Genet.* **21,** 20–24.

21. Mei, R., Galipeau, P. C., Prass, C., et al. (2000) Genome-wide detection of allelic imbalance using human SNPs and high-density DNA arrays. *Genome Res.* **10,** 1126–1137.

22. Stears, R. L., Martinsky, T., and Schena, M. (2003) Trends in microarray analysis. *Nat. Med.* **9,** 140–145.

23. Yershov, G., Barsky, V., Belgovskiy, A., et al. (1996) DNA analysis and diagnostics on oligonucleotide microchips. *Proc. Natl. Acad. Sci. USA* **93,** 4913–4918.

24. Hacia, J. G., Brody, L. C., Chee, M. S., Fodor, S. P., and Collins, F. S. (1996) Detection of heterozygous mutations in BRCA1 using high density oligonucleotide arrays and two-colour fluorescence analysis. *Nat. Genet.* **14,** 441–447.

25. Hacia, J. G., Woski, S. A., Fidanza, J., et al. (1998) Enhanced high density oligonucleotide array-based sequence analysis using modified nucleoside triphosphates. *Nucleic Acids Res.* **26,** 4975–4982.

26. Hacia, J. G., Edgemon, K., Fang, N., et al. (2000) Oligonucleotide microarray based detection of repetitive sequence changes. *Hum. Mutat.* **16,** 354–363.

27. Favis, R., Day, J. P., Gerry, N. P., Phelan, C., Narod, S., and Barany, F. (2000) Universal DNA array detection of small insertions and deletions in BRCA1 and BRCA2. *Nat. Biotechnol.* **18,** 561–564.

28. Kim, I. J., Kang, H. C., Park, J. H., et al. (2002) RET oligonucleotide microarray for the detection of RET mutations in multiple endocrine neoplasia type 2 syndromes. *Clin. Cancer Res.* **8,** 457–463.

29. Hacia, J. G., Sun, B., Hunt, N., et al. (1998) Strategies for mutational analysis of the large multiexon ATM gene using high-density oligonucleotide arrays. *Genome Res.* **8,** 1245–1258.

30. Primdahl, H., Wikman, F. P., von der Maase, H., Zhou, X. G., Wolf, H., and Orntoft, T. F. (2002) Allelic imbalances in human bladder cancer: genome-wide detection with high-density single-nucleotide polymorphism arrays. *J. Natl. Cancer Inst.* **94,** 216–223.

31. Hoque, M. O., Lee, C. C., Cairns, P., Schoenberg, M., and Sidransky, D. (2003) Genome-wide genetic characterization of bladder cancer: a comparison of high-density single-nucleotide polymorphism arrays and PCR-based microsatellite analysis. *Cancer Res.* **63,** 2216–2222.

32. Blohm, D. H. and Guiseppi-Elie, A. (2001) New developments in microarray technology. *Curr. Opin. Biotechnol.* **12,** 41–47.

33. Findlay, I., Matthews, P., and Quirke, P. (1998) Multiple genetic diagnoses from single cells using multiplex PCR: reliability and allele dropout. *Prenat. Diagn.* **18,** 1413–1421.

34. Wells, D., Sherlock, J. K., Handyside, A. H., and Delhanty, J. D. (1999) Detailed chromosomal and molecular genetic analysis of single cells by whole genome amplification and comparative genomic hybridisation. *Nucleic Acids Res.* **27,** 1214–1218.

35. Katz, M. G., Trounson, A. O., and Cram, D. S. (2002) DNA fingerprinting of sister blastomeres from human IVF embryos. *Hum. Reprod.* **17,** 752–759.

36. Piyamongkol, W., Bermudez, M. G., Harper, J. C., and Wells, D. (2003) Detailed investigation of factors influencing amplification efficiency and allele drop-out in single cell PCR: implications for preimplantation genetic diagnosis. *Mol. Hum. Reprod.* **9,** 411–420.

37. Harper, J. C. and Wells, D. (1999) Recent advances and future developments in PGD. *Prenat. Diagn.* **19,** 1193–1199.

38. Zhang, L., Cui, X., Schmitt, K., Hubert, R., Navidi, W., and Arnheim, N. (1992) Whole genome amplification from a single cell: implications for genetic analysis. *Proc. Natl. Acad. Sci. USA* **89,** 5847–5851.

39. Sermon, K., Lissens, W., Joris, H., Van Steirteghem, A., and Liebaers, I. (1996) Adaptation of the primer extension preamplification (PEP) reaction for preimplantation diagnosis: single blastomere analysis using short PEP protocols. *Mol. Hum. Reprod.* **2,** 209–212.

40. Dietmaier, W., Hartmann, A., Wallinger, S., et al. (1999) Multiple mutation analyses in single tumor cells with improved whole genome amplification. *Am. J. Pathol.* **154,** 83–95.

41. Cheung, V. G. and Nelson, S. F. (1996) Whole genome amplification using a degenerate oligonucleotide primer allows hundreds of genotypes to be performed on less than one nanogram of genomic DNA. *Proc. Natl. Acad. Sci. USA* **93,** 14,676–14,679.

42. Dean, F. B., Hosono, S., Fang, L., et al. (2002) Comprehensive human genome amplification using multiple displacement amplification. *Proc. Natl. Acad. Sci. USA* **99,** 5261–5266.

43. Salvado, C. S., Trounson, A. O., and Cram, D. S. (2004) Towards preimplantation diagnosis of cystic fibrosis using microarrays. *Reprod. Biomed. Online* **8,** 107–114.

44. Urakawa, H., El Fantroussi, S., Smidt, H., et al. (2003) Optimization of single-base-pair mismatch discrimination in oligonucleotide microarrays. *Appl. Environ. Microbiol.* **69,** 2848–2856.

45. Anthony, R. M., Schuitema, A. R., Chan, A. B., Boender, P. J., Klatser, P. R., and Oskam, L. (2003) Effect of secondary structure on single nucleotide polymorphism detection with a porous microarray matrix: implications for probe selection. *Biotechniques* **34,** 1082–1086, 1088, 1089.

46. Solinas-Toldo, S., Lampel, S., Stilgenbauer, S., et al. (1997) Matrix-based comparative genomic hybridization: biochips to screen for genomic imbalances. *Genes Chromosomes Cancer* **20,** 399–407.

47. Wong, K. K., Tsang, Y. T., Shen, J., et al. (2004) Allelic imbalance analysis by high-density single-nucleotide polymorphic allele (SNP) array with whole genome amplified DNA. *Nucleic Acids Res.* **32,** e69.

48. Dreesen, J. C., Jacobs, L. J., Bras, M., et al. (2000) Multiplex PCR of polymorphic markers flanking the CFTR gene: a general approach for preimplantation genetic diagnosis of cystic fibrosis. *Mol. Hum. Reprod.* **6,** 391–396.

49. Moutou, C., Gardes, N., and Viville, S. (2002) Multiplex PCR combining deltaF508 mutation and intragenic microsatellites of the CFTR gene for pre-implantation genetic diagnosis (PGD) of cystic fibrosis. *Eur. J. Hum. Genet.* **10,** 231–238.

50. Vrettou, C., Tzetis, M., Traeger-Synodinos, J., Palmer, G., and Kanavakis, E. (2002) Multiplex sequence variation detection throughout the CFTR gene appropriate for preimplantation genetic diagnosis in populations with heterogeneity of cystic fibrosis mutations. *Mol. Hum. Reprod.* **8,** 880–886.

51. Abou-Sleiman, P. M., Apessos, A., Harper, J. C., Serhal, P., Winston, R. M., and Delhanty, J. D. (2002) First application of preimplantation genetic diagnosis to neurofibromatosis type 2 (NF2). *Prenat. Diagn.* **22,** 519–524.

# Index